REDISCOVERING NARMADA VALLEY

ADIL KHAN

PAPER MISSILE

NIYOGI BOOKS

Published by
NIYOGI BOOKS
Block D, Building No. 77,
Okhla Industrial Area, Phase-I,
New Delhi-110 020, INDIA
Tel: 91-11-26816301, 26818960
Email: niyogibooks@gmail.com
Website: www.niyogibooksindia.com

Editor: Sukanya Sur
Design: Shashi Bhushan Prasad

ISBN: 978-93-89136-18-0
Publication: 2019

Printed at: Niyogi Offset Pvt. Ltd., New Delhi, India

CONTENTS

LIST OF ABBREVIATIONS

AIBP	:	Accelerated Irrigation Benefit Programme
ATR	:	Action Taken Report
CEA	:	Central Electricity Authority
CHPH	:	Canal Head Power House
CWINC	:	Central Waterways, Irrigation, and Navigation Commission
CWPC	:	Central Water and Power Commission
DPR	:	Detailed Project Report
EMC	:	Energy Management Centre
FCRA	:	Foreign Contribution (Regulation) Act
FRL	:	Full Reservoir Level
GRA	:	Grievance Redressal Authority
GSECL	:	Gujarat State Electricity Company Ltd.
IAS	:	Indian Administrative Service
KMCS	:	Khedut Mazdoor Chetna Sangath
kV	:	Kilovolt
kW	:	Kilowatt
MAF	:	Million Acre-Feet
MCM	:	Million Cubic Metre
MLD	:	Million Litre Daily
MoEF	:	Ministry of Environment, Forest, and Climate Change
MoU	:	Memorandum of Understanding
NABARD	:	National Bank for Agriculture and Rural Development
NBA	:	Narmada Bachao Andolan
NCA	:	Narmada Control Authority

NCB	:	Narmada Control Board
NHDC	:	Narmada Hydroelectric Development Corporation Ltd.
NHPC	:	National Hydroelectric Power Corporation
NIMR	:	National Institute of Malaria Research
NKS	:	Narmada-Kshipra-Simhastha
NPG	:	Narmada Planning Group
NRCD	:	National River Conservation Directorate
NRHM	:	National Rural Health Mission
NSSO	:	National Sample Survey Organization
NVDD	:	Narmada Valley Development Department
NVDA	:	Narmada Valley Development Authority
NVDP	:	Narmada Valley Development Project
NWDT	:	Narmada Water Disputes Tribunal
NWDTA	:	Narmada Water Disputes Tribunal Award
NWRDC	:	Narmada Water Resources Development Committee
PESA Act	:	Panchayats (Extension to Scheduled Areas) Act
PIL	:	Public Interest Litigation
PPP	:	Public-Private Partnership
R&R	:	Rehabilitation and Resettlement
RBPH	:	River Bed Power House
RCNCA	:	Review Committee of NCA
SLP	:	Special Leave Petition
SOI	:	Survey of India
SSDP	:	Sardar Sarovar Development Plan
SSCAC	:	Sardar Sarovar Construction Advisory Committee
SSPAC	:	Sardar Sarovar Project Advisory Committee
TAP	:	Tribal Action Plan
WAPCOS	:	Water and Power Consultancy Services
WRD	:	Water Resources Department

LIST OF TABLES

Disclaimer: All tables developed by the author based on data from the Narmada Valley Development Authority.

FOREWORD

Narmada had been on the minds of visionaries of independent India, who observed immense potential in this river. Pandit Jawaharlal Nehru and Sardar Vallabhbhai Patel, the first prime minister and home minister, respectively, of independent India, were pioneers of the plan to utilize Narmada's water for agricultural development.

With the passage of time, the idea of utilizing Narmada's water gained ground, but later on transformed in to a dispute over sharing the water among the states of Madhya Pradesh, Gujarat, and Maharashtra. When an amicable solution could not be reached, the then prime minister of India Mrs Indira Gandhi constituted the NWDT in 1969. The tribunal declared its final award after a decade-long exercise. However, when the party states initiated action to follow the directions of the award, people started opposing the construction of big dams. The scattered opposition was then organized and effectively represented by the movement NBA. Meanwhile, the governments of Madhya Pradesh, Gujarat, and Maharashtra indulged in the politics of water by making political statements and counter statements.

Over a period of time, the construction of major irrigation projects was initiated, but faced enormous challenges of human displacement and resettlement. In one way or the other, the story is repeated still with the upcoming irrigation projects of the valley.

For a long time, there has been a need to rediscover and disseminate all aspects of the subject in a systematic, authentic, and usable form to make people aware about all the concerned aspects. The eventful and kaleidoscopic story of Narmada is worth knowing for the new generation from the past, present, and future perspectives. This book is an endeavour to achieve the same.

The author of this book, Adil Khan, has been associated with Narmada-related activities for about one and a half decades. As the Joint Director, Public Relations, he has been working in association with the decision-making political and administrative authorities, irrigation engineers, agriculture experts, social workers, and people of the Narmada Valley.

The author, of his own interest, has widely travelled in the valley. Many of his articles on the Narmada-related issues have been published in leading newspapers. This book is an excellent outcome of his prowess in dealing with the issue. He has treated the subject in an interesting way and made it highly readable.

I am confident that this book will be welcomed.

Aditya Vijay Singh
Former Chief Secretary, Govt. of Madhya Pradesh
Former Chairman, NVDA
Former Chief Election Commissioner, MP

PREFACE

The river Narmada has been a subject of wide public and political discussion since the dispute over sharing its water surfaced between Gujarat, Madhya Pradesh, and Maharashtra. When these states could not reach any agreement, the government of India constituted Narmada Water Dispute Tribunal (NWDT) in 1969 to adjudicate the dispute. After a decade-long exercise, the tribunal gave its final award in December 1979, settling the contention over the sharing of water and other related interstate issues.

After overcoming the preliminary problems, the states of Gujarat and Madhya Pradesh initiated construction of dams, but were met with a feverish pace of public opposition. In due course of time, people's opposition to big dams was streamlined and defined by Narmada Bachao Andolan (NBA). This movement registered effective protests, rallies, and dharnas under the leadership of Medha Patkar. The opposition was based on the plea that dam reservoirs would engulf large areas of productive lands, displacing lakhs of people. Moreover, the construction of large dams would cause permanent harm to the forest and the surroundings, leading to loss of biodiversity. The issues raised by the NBA were serious. Many NGOs in the country and across the world inclined to extend their support to the cause. Thus, Narmada and the NBA became global in extent.

While the construction of dams took on a very slow pace, the country started opening its doors to the trends of globalization. Soon, the effects of a new economic order had started trickling in both urban and rural areas. In the changing atmosphere, farmers started aspiring about multicropping agriculture in order to enhance their economic stature. The younger farming generations began to change their minds by admitting that multicropping practices would need efficient flow irrigation, which could be guaranteed only by big dam projects.

As the construction of dams attained some pace, the challenges accompanying it multiplied simultaneously. The displacement and resettlement of a large population coming under submergence, the compensation for submerged agriculture, lands, and houses, and the conservation of forests and environment were some of the hard-pressing challenges to overcome. The concerned states made efforts to meet these challenges and formulated their own strategies to move forward.

As the Joint Director, Public Relations, Government of Madhya Pradesh, for Narmada-related matters, I realized that it was difficult to get a holistic view of Narmada and the related issues in order to come to terms with the realities on ground. The various issues that originated have intermingled with each other, making the whole subject complicated. Many times, a state of confusion has arisen and misplaced information have spread in public, political, and media circles. However, a subject as important as this has been altogether absent from the school and college-level curriculum, leaving the young generation unaware about the crucial activities going on in Narmada Valley.

Hence, I decided to act upon the need to rediscover all the issues related to Narmada Valley and put them in a systematic, unbiased, simple, and chronological manner. I also feel that a comprehensive account of the Narmada water dispute may help in some way to deal with the future river water disputes, if any, in our country.

I hope that my attempt shall fulfil the need of getting an integrated view of Narmada and Narmada Valley in its comprehensiveness.

THE RIVER AND ITS FAMILY

Narmada is the fifth largest river in the Indian subcontinent. After Godavari and Krishna, it is the third longest river that flows through India. Narmada, considered to be one of the holiest rivers of the country, occupies a place of pride in the religious beliefs and ethos of Indian people. 'Narmada' is a Sanskrit word, meaning 'the giver of pleasure'. It is almost like a manifest life, providing as a mother does to the populace residing in its hinterland. The forests spread over its vast area, and the variety of wildlife and chirping birds seen wandering in these forests provide an incomparable natural beauty to Narmada Valley. Be it wildlife, birds, tribes subsisting on forests, or cultivators growing foodgrains on lands, Narmada catchment provides shelter and bounty to all who seek refuge in it. Viewed in this aspect, the river Narmada and the valley constitute a peerless meeting ground of nature and diversities of human existence.

Narmada originates from Amarkantak, a hallowed place ensconced in the Maikal Hills spread across the district of Anuppur in Madhya Pradesh. Amarkantak is at a height of 1,051 metres (3,468 ft) above the sea level. Starting from Amarkantak, Narmada flows over a length of 1,312 kms before falling into the Gulf of Khambhat in Gujarat. It covers a distance of about 1,077 kms in Madhya Pradesh, flowing from its origin through the districts of Anuppur, Dindori, Mandla, Jabalpur, Narsingpur, Hoshangabad, Harda, Khandwa, Khargone, and Barwani. Thereafter, it flows for a distance of 35 kms along the border of Madhya Pradesh and Maharashtra, and a further stretch of 39 kms along the border of Maharashtra and Gujarat. It further flows for the remaining 161 kms over the districts of Narmada and Bharuch of Gujarat before falling into the Gulf of Khambhat, nearly 50 kms away from Bharuch, finally to mingle with the Arabian Sea. The Narmada makes estuaries when it unites with the Arabian Sea. It flows from east to west, and is the only major river flowing towards the west in the Indian subcontinent.

THE RIVER RECEIVING WATER FROM LARGE AREAS

The total catchment area of Narmada is 98,796 sq. kms of which 85,859 sq. kms (87%) falls in Madhya Pradesh. Out of this total catchment area, nearly 1,538 sq. kms and 11,399 sq. kms fall in Maharashtra and Gujarat respectively. The average rainfall in the catchment areas is 112 cms. The catchment area of Narmada covers 22 districts of Madhya Pradesh, 2 districts of Chhattisgarh, 1 district of Maharashtra, and 5 districts of Gujarat, which is mentioned as follows.

Madhya Pradesh	:	Anuppur, Shahdol, Mandla, Dindori, Balaghat, Seoni, Jabalpur, Narsingpur, Sagar, Damoh, Chhindwara, Hoshangabad, Betul, Raisen, Sehore, Khandwa, Khargone, Barwani, Dewas, Indore, Dhar, Alirajpur
Chhattisgarh	:	Rajnandgaon, Kawardha
Maharashtra	:	Dhule
Gujarat	:	Panchmahal, Vadodara, Narmada, Bharuch, Surat

TRIBUTARIES OF NARMADA

Although there are a number of tributaries of Narmada, the main ones whose individual catchment areas exceed 500 sq. kms number to 41. Of these, 19 tributaries meet the Narmada on the right bank, whereas 22 tributaries join it on the left bank. Again, of these 41 main tributaries, 39 lie in Madhya Pradesh and only 2 join them in Gujarat. The details of these tributaries have been listed as follows:

Tributaries of Narmada

State	Tributaries on the Right Bank		Tributaries on the Left Bank	
Madhya	Silgi	Jamner	Kharmer	Kundi
Pradesh	Balai	Chankeshar	Dudhi	Soner
	Gaur	Khari	Chota Tawa	Ganjal
	Hiran	Kanar	Burhner	Borad
	Biranj	Choral	Sukhari	Sher
	Tendoni	Karam	Kaveri	Ajnal
	Barna	Maan	Banjar	Deb
	Kolar	Uri	Tawa	Shakkar
	Sip	Hathni	Kharkiya	Machak
			Temur	Goi
			Hather	
Gujarat	Orsang		Karjan	

Besides these, there are a large number of sub-tributaries of Narmada, the small rivers and rivulets which join the tributaries that later join Narmada. Thus, Narmada is a large family with hundreds of member tributaries, sub-tributaries, and rivulets spread over a large area in northern India.

IDEA FOR ARRESTING NARMADA FLOW

The idea for arresting the water flow by constructing a dam on Narmada for the purpose of irrigation was conceived towards the end of the nineteenth century under British colonial rule. The first Irrigation Water Commission of India in its report, submitted in the year 1901, had underlined the need for constructing a barrage on Narmada near Bharuch (now in Gujarat). During 1946, the then state of Bombay and the Central Provinces and Berar had jointly requested the Central Waterways, Irrigation, and Navigation Commission (CWINC), presently known as Central Water Commission (CWC), that research studies pertaining to flood control, navigation, irrigation, and generation of hydroelectricity with respect to Narmada be undertaken.

Based on the said proposal, the government of India constituted an ad hoc committee under the chairmanship of A.N. Khosla, then chairman of CWINC. The committee headed by Khosla recommended in the year 1948 that a detailed survey and studies be conducted for projects at Bargi, Tawa, Punasa [presently called Indira Sagar Project (ISP)], and Bharuch [presently called Sardar Sarovar Project (SSP)]. It is to be noted here that the idea of development through big dams was highly inspired by the vision of the first prime minister of independent India Pandit Jawaharlal Nehru, who was in favour of the construction of large dams and called the gigantic Koyna and Bhakra Nangal dams the modern centres of pilgrimage.

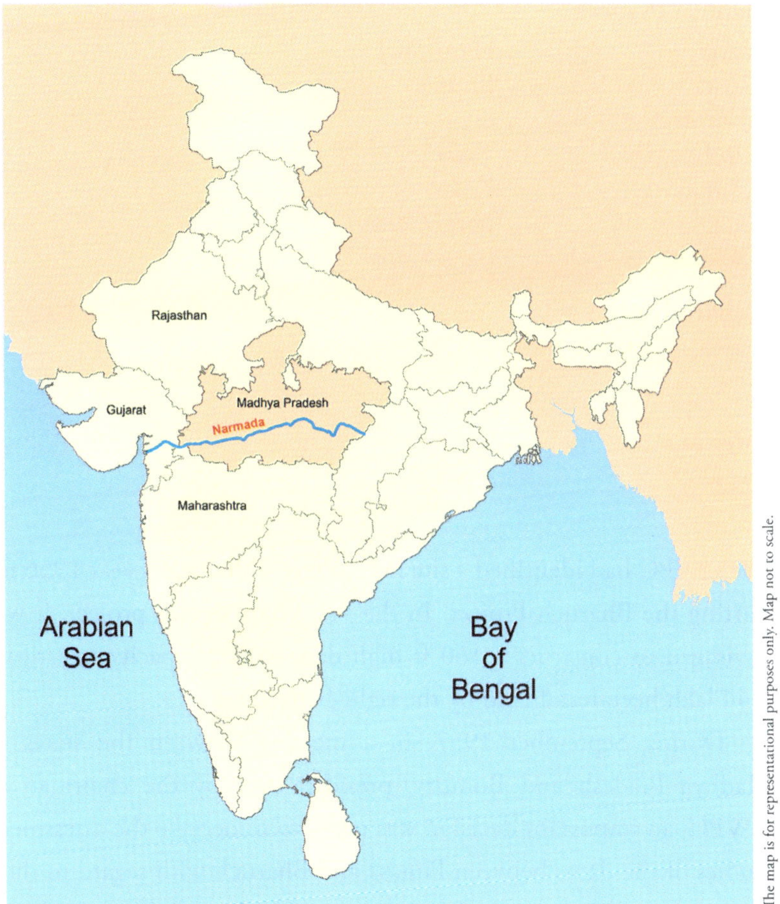

The Narmada—Lifeline of Gujarat and Madhya Pradesh

In the year 1955, the CWINC was renamed as Central Water and Power Commission (CWPC). This commission, after studying the potential of generation of hydroelectricity of Narmada, suggested the possibility of setting up hydroelectricity units at 16 places on Narmada and its tributaries. According to preliminary estimates of the study, it was possible to generate about 1,300 megawatt of power at these units. The study identified 14 places in Madhya Pradesh, 1 in Maharashtra, and 1 in Gujarat, which are mentioned as follows:

Madhya Pradesh

Bargi	Harinphal
Rosra	Budner
Basania	Sitareva
Hoshangabad	Tawa
Chinki	Kolar
Punasa	Bargi Canal I
Barwaha	Bargi Canal II

Maharashtra

Keli (Jalsindhi)

Gujarat

Gora

The CWPC had identified a site near village Gora in the year 1956 for starting the Bharuch Project. In the first phase of this project, it was envisaged to construct a 160 ft high dam with a capacity to irrigate 4.44 lakh hectares of land by the right-bank canal.

During September 1957, in a meeting between the states of Madhya Pradesh and Bombay, presided over by the chairman of CWPC, an important decision was taken to undertake the assessment studies of the dams between Punasa and Bharuch with regard to their potential for irrigation and generation of hydroelectricity. Both the states also agreed to equally share the costs involved.

Dr K.L. Rao, then member of the CWPC, in February 1957, inspected the site of Bharuch Project near village Gora. After inspecting, he suggested another site, about 2.5 kms upstream of the proposed dam, near Navagam village. Based on the recommendations of Dr Rao for the appropriateness of the place, the dam was decided to be set near Navagam village. Later, in the year 1959, the CWPC proposed construction of the dam at the new site that was near Navagam (now known as Sardar Sarovar Project) in two phases.

In January 1959, the Ministry of Irrigation and Power constituted a panel of experts to revise the Bharuch Irrigation Project. The panel, integrating the two-phase construction of the project, recommended the construction of a 320 ft high dam. The panel also observed that the command area could be suitably augmented by the construction of a high-level canal network in north Gujarat and the Rann of Kutch. In the wake of Bombay State Regulation Act, Bombay was divided into two parts and, on 1 May 1960, two new states, Gujarat and Maharashtra, came into existence. Since the proposed site of the Navagam Dam fell within the jurisdiction of Gujarat, all work pertaining to the construction of the said project was transferred to Gujarat.

During August 1960, the Planning Commission of India approved the construction of Bharuch Dam in two phases. The first phase would include the construction of the dam up to a height of 162 ft and the construction of a canal network at the lower level, thereby building a capacity of irrigating about 3.89 lakh hectares of land. During the second phase of construction, the height of this dam would be raised up to 320 ft and the construction of a network of high-level canal system would be undertaken, so that an additional capacity of irrigating 3.64 lakh hectares of land is created. In addition, it was also approved that, upstream of this dam, dams at Punasa (now known as Indira Sagar Project) and other sites be constructed so that after meeting the irrigation requirement for Madhya Pradesh, 625 megawatt of hydroelectricity could be generated by regulated discharge of water.

Following the consent given by the Planning Commission, the government of Gujarat accorded administrative approval in February 1961 to the construction of the first phase of Bharuch Dam. Pandit Jawaharlal Nehru, then prime minister of India, laid the foundation of this dam near Navagam in Gujarat on 5 April 1961.

STRUGGLE FOR A GREATER SHARE OF NARMADA

The states of Gujarat and Madhya Pradesh tried their best to get the maximum of Narmada water. This led to a dispute between the two states. Traces of this dispute go back to the decade of the 1960s.

In the years 1961–62, a detailed survey was carried out of the area which came under submergence due to Hiranphal Dam in Madhya Pradesh and Keli (Jalsindhi) Dam on the border of Madhya Pradesh and Maharashtra. This work was entrusted to the Survey of India (SOI) by the CWPC. The SOI submitted reservoir sheets of the two proposed dams to the commission. These sheets contained details of the submergence area due to the proposed dam reservoirs.

Gujarat swung into action on the basis of data provided in the reservoir sheets. It conducted geographical and geological studies pertaining to the Navagam Dam and decided upon the construction of a larger dam. This dam site was chosen at a site about 4.0 metres upstream of the previously proposed area. This decision of Gujarat led to the possibility of drowning the proposed Hiranphal and Keli (Jalsindhi) dams. It was the beginning of a conflict regarding the construction of dams in the lower Narmada region.

In the year 1963, the then irrigation minister Dr K.L. Rao advised to construct Navagam Dam at the proposed new site of the Bharuch Project. Gujarat wanted to construct a taller dam at the site, to which Dr Rao, who was also a leading irrigation engineer, advised to restrict the height to 425 ft because a larger dam would cause increased submergence. It was also proposed that the hydroelectricity generated here would be made available to Gujarat. It was further advanced that the power generated at the proposed Punasa Dam (now called Indira Sagar Dam) in Madhya Pradesh would be shared equally between the three states of Madhya Pradesh, Maharashtra, and Gujarat.

THE RENEGED AGREEMENT

In November 1963, a meeting between the chief ministers of Madhya Pradesh and Gujarat was held in Bhopal under the chairmanship of the Central Minister for Irrigation and Power on the issue of the disputed height of Navagam Dam. A draft prepared in Delhi for resolving the dispute was discussed in the meantime. After the discussion, a broad agreement on the related issues was reached. This was known as the Bhopal Agreement. The salient points of this agreement were:

- Gujarat will construct a 425 ft high dam at Navagam, and all benefits accruing from there shall go to Gujarat.
- The height of Punasa Dam in Madhya Pradesh shall be 850 ft. The cost of construction and hydroelectricity generated would be shared between Gujarat and Madhya Pradesh in the ratio of 1:2.
- Half of the power share of Madhya Pradesh shall be provided to Maharashtra for a period of 25 years and in lieu, thereof, the state of Maharashtra shall make available to Madhya Pradesh one-third of the cost of construction of Punasa Dam in the form of loan. Madhya Pradesh shall repay this loan within a period of 25 years.
- The Bargi Dam shall be constructed in the first phase up to a height of 1,365 ft and, in the second phase, raise it to the height of 1,390 ft. For this purpose, the states of Gujarat and Maharashtra shall make available to Madhya Pradesh a loan of Rs. 10 crore.

Later, Madhya Pradesh reneged on the Bhopal Agreement. D.P. Mishra, then chief minister of Madhya Pradesh, wrote to Dr Rao on 28 November 1963, stating that the Bhopal Agreement was not acceptable. However, the state of Gujarat, in furtherance of the Bhopal Agreement, prepared a project report of the Navagam Dam (with a height of 425 ft) and submitted it to CWPC on 14 February 1964. Madhya Pradesh took serious objection to keeping the height of Navagam Dam at 425 ft as per the Bhopal Agreement. It made the plea that since the level of the Narmada riverbed on the border of Madhya Pradesh and Gujarat was 162 ft, the Navagam Dam's height should not be more than that.

GOVERNMENT'S ACTION ON THE DISPUTE

In the wake of the rejection of the Bhopal Agreement by Madhya Pradesh, Government of India constituted a high-power committee on 5 September 1964 called the Narmada Water Resources Development Committee (NWRDC) under the chairmanship of Dr A.N. Khosla, then governor of Orissa. The committee included many senior civil engineers.

The Khosla Committee submitted its report on 1 September 1965 to the government of India. A master plan for the development of Narmada water resources was presented in this report. It was recommended that 12 identified major dams should be built in Madhya Pradesh, and that the height of the Navagam Dam in Gujarat should be 500 ft. It was assessed that Navagam Dam could produce 1,400 megawatt hydroelectricity. The Khosla Committee also proposed equitable distribution of water amongst the related states. It was suggested that out of 30.40 Million Acre-Feet (MAF) water available in Narmada, 15.6 MAF should go to Madhya Pradesh, while Gujarat and Maharashtra should get 10.65 MAF and 0.10 MAF water respectively. It was also recommended that 0.25 MAF water should be allocated to Rajasthan, even though it had not made any claims on Narmada. The remaining 3.9 MAF water would be reserved for the generation of hydroelectricity.

Even though Gujarat and Rajasthan accepted the recommendations of the Khosla Committee, Madhya Pradesh and Maharashtra rejected them outright. The argument of Madhya Pradesh was that since no land of Rajasthan was part of the Narmada basin, it was not entitled to use its water. Madhya Pradesh also claimed cent percent right on the hydroelectricity that was to be generated in its boundary. Thus, Narmada became a subject of complex dispute between Gujarat, Rajasthan, Madhya Pradesh, and Maharashtra.

In order to reach a consensus amongst the four states with regard to the recommendations of the Khosla Committee, the Minister for Irrigation and Power made visits to the four states. During these visits in May–June 1966, he discussed the matter with the respective chief ministers. Later, during July–August 1966, the CWC organized a meeting of the officers of these states to discuss technical issues contained in the report of the Khosla Committee.

A DEAL BYPASSING GOVERNMENT OF INDIA

While the Khosla Committee was still engaged in resolving the Narmada water dispute, Madhya Pradesh and Maharashtra struck a deal for constructing a dam on Narmada in a place called Jalsindhi. Jalsindhi is located between Hiranphal and Navagam dams (presently Sardar Sarovar Project) on the border of Madhya Pradesh and Maharashtra. Under the terms of the agreement, signed on 5 April 1965, Maharashtra agreed to conduct necessary research and prepared a Detailed Project Report (DPR) for the generation of hydroelectricity at Jalsindhi Dam in the context of the master plan of Madhya Pradesh. While Madhya Pradesh would acquire land and extend necessary facilities, Maharashtra would take up construction of the dam, powerhouse, and other related works. The cost of the project was to be shared between the two states in the same ratio as the use of the river in each state. This agreement came to be known as the Jalsindhi Agreement. However, the Khosla Committee rejected it by saying that the 'agreement is based on selfish motives'. The government also overlooked this agreement and, thus, no action could be taken on the terms of the deal.

In August 1966, a meeting of the chief ministers of Gujarat, Rajasthan, Maharashtra, and Madhya Pradesh was organized under the chairmanship of the Central Minister for Irrigation and Power to discuss the Narmada issue. Acting on the decision taken in this meeting, the chief ministers of Madhya Pradesh and Gujarat met in May 1967 at Pachmarhi to resolve the dispute. Thereafter, a meeting was organized under the chairmanship of the CWPC in June 1967 at New Delhi between the officers of the states of Madhya Pradesh and Gujarat. This was followed by yet another meeting of the chief ministers of the four states under the chairmanship of the Central Irrigation Minister in December 1967 at New Delhi. But, despite these efforts, Narmada water dispute could not be resolved.

TRIBUNAL SET UP BY PRIME MINISTER

The issue of the height of the dam at Navagam led to a dispute regarding the use of Narmada water among Madhya Pradesh, Gujarat, and Maharashtra. In this connection, Gujarat lodged a complaint in July 1968 under the Interstate River Water Disputes Act, 1956. Acting upon the complaint, Prime Minister Indira Gandhi directed to constitute a tribunal to look into the matter and resolve the issue. Accordingly, Government of India set up a tribunal under Section 4(1) of the Water Disputes Act. The tribunal (*panchat*), comprising three judges for the redressal of the complaint by Gujarat, came in to existence on 6 October 1969. The tribunal came to be known as the Narmada Water Disputes Tribunal (NWDT). Later, the government included certain issues raised by Rajasthan within the terms of reference of the tribunal. Justice V. Ramaswami of the Supreme Court was appointed the chairman of this tribunal. Justice A.K. Sinha of the Calcutta High Court and Justice M.R.A. Ansari of Jammu and Kashmir High Court were appointed members of the tribunal. Following transfers of the judges during the proceedings of the tribunal, the government appointed three more members, Justice G.S. Mathur, Allahabad High Court; Justice E. Venkatesan, Andhra Pradesh High Court; and Justice V.P. Gopalan Nambiar, Kerala High Court.

The NWDT, in order to conduct its proceedings, availed of the services of the following experienced and competent technical experts (assessors):

- Dr M.R. Chopra, former chairman of CWPC and ex vice chancellor, Roorkee University
- C.S. Padmanabha Iyer, chief engineer, Kerala
- Balwant Singh Nag, former advisor, Planning Commission, India
- Dr Ambika Singh, assistant director general, Indian Council of Agricultural Research
- Dr M.B. Hukkeri, assistant director general, Indian Council of Agricultural Research

OBJECTION OF THE JURISDICTION OF THE TRIBUNAL

In July 1972, the NWDT decided on a number of legal issues involved. The objection by Madhya Pradesh was accepted on the grounds that since no part of Rajasthan fell within the Narmada basin, it was not entitled to lay any claim on its water. The tribunal, however, rejected the plea of Madhya Pradesh stating that since the land of Madhya Pradesh and Maharashtra would come under submergence due to the construction of the dam at Navagam, it would not be a water dispute but a land dispute. The view of Madhya Pradesh was that since the NWDT was constituted under the provision of the Interstate Water Disputes Act, it had no jurisdiction to adjudicate on the land dispute.

The tribunal held the view that the main issue regarding the water dispute was the 'equitable apportionment of water among states', and that the height of the dam would be based on this apportionment. So the decision of the tribunal fell within the statutory rules governing the 'equitable apportionment' of water. Hence, the tribunal was competent to give directions to Madhya Pradesh and Maharashtra for making land available for construction of the Navagam Dam. In the year 1972, Madhya Pradesh submitted a master plan before the tribunal for the development of water resources of Narmada.

ATTEMPT TO WITHDRAW FROM THE TRIBUNAL

Madhya Pradesh and Rajasthan filed an appeal before the Supreme Court against the said legal decision passed by the NWDT. The Supreme Court granted 'stay' on this appeal. On 22 July 1972, the states of Madhya Pradesh, Gujarat, Maharashtra, and Rajasthan came to an agreement that 0.5 MAF water of Narmada shall be made available to Rajasthan. It was also agreed that the Narmada water dispute would be withdrawn from the NWDT and entrusted to the prime minister for its resolution. As per this agreement, the four states entrusted the Narmada water dispute to then Prime Minister Indira Gandhi for making the verdict. The matter, however, remained pending with the prime minister for two years, and no decision could be taken thereon.

RETURN TO THE TRIBUNAL OWING TO INDECISIVENESS

After waiting for two years, on 12 July 1974, the four states once again came to an agreement to entrust the Narmada water dispute back to the tribunal. Following this agreement, Madhya Pradesh and Rajasthan withdrew the appeals filed by them before the Supreme Court. The tribunal again started hearing pleas and arguments of the four states regarding their stands on the dispute. The Narmada water dispute is a unique interstate dispute of India in the sense that proceedings to resolve the dispute continued for 10 years. In the process, dozens of legal luminaries and technical experts representing the states appeared before the tribunal.

Such extensive and deep analysis of the legal and technical aspects of any river dispute was never done in the past. Another aspect worth mentioning here is that the final verdict passed by the tribunal on the water dispute was wholly accepted by the four states without any reservations whatsoever, and they set about complying with the same.

The tribunal extensively toured different areas of the four states and did spot inspection of sensitive sites. A decade-long study, analysis, and research of all the related issues and aspects of the dispute yielded in the form of an extensive report of conclusions running in four volumes. The report was issued on 16 August 1978. One member of the tribunal gave his independent views on certain matters. The report was issued by the majority of the chairman and one member. After analysing and examining the respective claims of the four states at length, the NWDT held that the height of the Navagam Dam should be 455 ft and issued instructions regarding the allocation of Narmada water to the four states. The central government and the four states submitted applications (references) before the NWDT, seeking clarification on certain points.

Gujarat, in its reference, sought clarification on various issues and made a few suggestions as well. The main suggestions submitted by Gujarat were as follows:

- The height of Navagam Dam should be raised to 473 ft.
- The foundation of the dam should be permitted to be so

constructed as to sustain a dam height of 530 ft, so that in future, if required, the dam height can be easily raised.
- Gujarat should also get a share in the hydroelectricity to be generated at Indira Sagar Dam in Madhya Pradesh.

The main suggestions contained in the reference made by Madhya Pradesh may be summarized as follows:
- The height of Navagam Dam should be 443 ft.
- In order to utilize the allocated water, permission should be granted to use it as per convenience both within and outside the basin.

Maharashtra and Rajasthan in their references sought some clarifications. Likewise, the Central Government also filed petitions for clarifications on certain issues.

LIGHTER MOMENTS DURING TRIBUNAL PROCEEDINGS

The NWDT was constituted in the year 1969. Hearing before the tribunal on behalf of the states of Madhya Pradesh, Gujarat, Maharashtra, and Rajasthan continued for 10 long years. The proceedings of the tribunal at New Delhi used to be conducted in Vigyan Bhawan. All the states engaged top-class counsels/lawyers to plead their cases before the tribunal. A team of senior engineers and planners would brief these counsels/lawyers on technical matters and provide other necessary details and documents. The then judge of the Supreme Court, Justice V. Ramaswami, was the chairman of the tribunal; the members thereof were judges of the high courts of various states. Thus, the proceedings of the NWDT, aimed at resolving the Narmada water dispute, were highly technical and legalistic in nature. During the proceedings, issues ranging from geology, geography, soil chemistry, minutiae of agricultural aspects, hydrology, demography, and social and economic aspects were discussed and argued.

Despite the discussions being serious, complex, and, in a way, insipid during the proceedings of the tribunal, a few moments of humour provided a welcome relief from the sombre mood of the arguments. The people who were witness to the proceedings of the tribunal still remember few such incidents, some of which are narrated here.

Senior advocate N.K. Palkiwala, appearing on behalf of Madhya Pradesh, observed before the tribunal that Gujarat had stated in its complaint that Jalsindhi and Hiranphal projects planned on the upstream of Navagam (Sardar Sarovar) Dam were hydroelectric ventures. Thus, these projects did not appropriate any water to be of consumptive use. Though neither of the dams diminished the share for Gujarat, Gujarat still opposed the construction of these two projects, claiming that the Sardar Sarovar Dam could be raised to enable Gujarat to utilize its equitable share for irrigation and power generation. The proposal of Gujarat would submerge large tracts of lands without the consent of the states of Madhya Pradesh and Maharashtra, which was not permissible.

On this, the chairman of the tribunal, Justice Ramaswami said, 'Palkhiwala, you have agreed to submerge your territory by the Hirakud Project of Orissa. You have agreed to submerge your territory by the Rihand Dam in Uttar Pradesh, and you have agreed to submerge your territory by another Matatila Dam on the Betwa River in Uttar Pradesh. On this analogy, if a decision is taken to let your land be submerged by Sardar Sarovar Dam, what objection can you possibly have to this?'*

Palkhiwala readily retorted, 'My lord, I can give my hat to a neighbour, but he cannot go to the court of law and demand my hat.'

Reacting to Palkhiwala's statement, the chairman of the tribunal observed that Gujarat was willing to pay Madhya Pradesh for the

* All the dialogues in this section are transcriptions of the author's conversations with the octogenarian irrigation engineer L.K. Wagh in 2010.

land that may be submerged due to Sardar Sarovar Dam. This elicited a more stirring response from Palkhiwala, 'If Gujarat, thus, continues to buy lands of neighbouring states, what will happen to these states? Three lakh acres of land of Madhya Pradesh comes under submergence of Sardar Sarovar Dam, and the state of Madhya Pradesh has only so much land. In that eventuality, Madhya Pradesh will just vanish.' To this, the chairman, members, and other advocates burst into roaring laughter.

J.M. Thakre, Advocate General of Gujarat, wanted to read out four lines from the decision of Krishna Water Dispute Tribunal in support of his arguments. At this stage, the chairman of the tribunal observed that there was no need to refer to the decision of another river dispute. Thereafter, he asked other members of the tribunal if they should take into account the decision of Krishna Water Dispute. At this, Justice Nambiar, judge of Kerala High Court and member of the tribunal, remarked, 'Light should be welcomed from wherever it comes.'

The chairman then smilingly permitted Thakre to read out the relevant lines.

On one occasion, the advocate general of Gujarat was quoting from the Khosla Committee Report of 1965. The chairman intervened to say that Dr Khosla had the habit of taking decisions in advance and submitting arguments and reasons in support later. This observation evoked instantaneous silence in the hall. The chairman Justice V. Ramaswami, sensing the tension caused by his remark, and with a view to detoxify his remark, further observed, 'I also do the same.' This lightened the atmosphere at once, making every one have a hearty laugh.

During the course of hearing, Justice M.R.A. Ansari, member of the tribunal, said that each state was blaming the other for misguiding the tribunal. 'We have to see who finally succeeds in doing that.'

K.K. Jain, a senior advocate from Rajasthan, observed that the entire basin of Narmada would cause speedy silting of all the dams on Narmada. The proposed dam at Jalsindhi on the border of Madhya Pradesh and Maharashtra would also be full of silt in a short time. The chairman inquired as to how many years it would take for Jalsindhi Dam to be silted. Jain readily answered, 'Two to three years.' Everyone burst into laughter at this. The chairman, assuring seeming seriousness on the ridiculous reply of the senior advocate of Rajasthan, observed, 'Okay, proceed further. Your notes are excellent.'

When Justice M.R.A. Ansari, then chief justice of Jammu and Kashmir High Court, joined the tribunal as a member, the chairman, during proceedings, observed that Ansari would like to ask the concerned states a few questions. 'But he would first go through the written notes submitted by you. It might happen that thereafter he would have no questions to ask.'

The senior advocate of Madhya Pradesh Dr Y.S. Chitale was arguing, inter alia, that the information submitted by Gujarat on the irrigation classification of the soil in the command area of Sardar Sarovar was one notch higher than actual. The soil classified as number one actually ranked two, number two as three, and so on. He questioned the claim of water by Gujarat. The tribunal questioned as to why a state should suffer for the mistake of a few of its officers. Then Dr Chitale started analysing the soil of Madhya Pradesh. Dr M.R. Chopra, assessor of the tribunal, expressed that the pH values of the soil of Madhya Pradesh were many times more than 8 or 8.5, and sometimes 9, but have shown 7 and 7.5 to be the permissible range.

Dr Chopra had made submissions on many occasions, breaking the flow of arguments made by the advocate of Madhya Pradesh. On this, the chairman said, 'Dr Chopra, why do you stand up during the flow of arguments and try to reply taking Gujarat's side?' Dr Chopra replied, 'I have to bring the facts to the notice of the tribunal. Otherwise, what should I do?' The chairman answered, 'You have to keep your mouth shut.' Addressing the secretary of the tribunal, he asked him to get a book on soil from his chamber. After getting the book, the chairman asked Vijay Bahadur Singh (Director of Agriculture, Government of Madhya Pradesh) to read the passage on the pH value. It said that the pH value of soil depends upon the season and timing, when, in the rainy season, the value of pH works out to be higher than normal. He suggested, therefore, that it was best not to take into account the pH values while classifying soil. The chairman asked Dr Chopra, 'What do you say now?' Ganesh, an advocate assisting Dr Chitale, confided to his colleague Advocate Shekhar Bhargava that it was clear from the proceedings that whenever Dr Chopra said something, the chairman put his foot in it.

Generally, advocates have the habit of profusely giving hands to the judges or expressing obligation on the slightest inclination of the judges towards their viewpoint. Use of words such as 'I am much obliged', 'I am grateful' were common. Every afternoon before the lunch break, the chairman used to say, 'Let us stop here,' and the advocate general of Gujarat would reply, 'As is convenient to your lordships.'

On the last day of the hearing on 15 March 1978, the chairman announced that there will be no further hearings. From force of habit, the advocate for Maharashtra replied, 'We are much obliged.' Once again, laughter rippled through the room.

HISTORIC VERDICT OF THE TRIBUNAL

Hearing of petitions continued for about one year and, finally, the tribunal rejected all the suggestions submitted by Gujarat. It also rejected the suggestion of Madhya Pradesh who wanted the height of the dam to be kept at 443 ft. But the demand of the latter saying that the allocated water might be permitted to be utilized as per convenience both inside as well as outside the basin was conceded. After due consideration of the petitions for clarification, the NWDT in December 1979 finally passed its historic order known as the Narmada Water Disputes Tribunal Award (NWDTA). After the gazette notification by Government of India on 12 December 1979, the award came into existence officially and soon was binding on all the four states. As per provisions laid down in Sections 6 and 11 of the Interstate River Water Disputes Act 1956, the award of the NWDT is binding on the four states, and it cannot be challenged before any court.

THE GIST OF THE FINAL ORDER AND DECISION OF THE TRIBUNAL

The final order of the tribunal runs in 16 parts (volumes), the sum and substance whereof are as follows:

Clause I: The orders of the tribunal shall come into operation from the date of its publication in the official gazette under Section 6 of Inter State Water Deputes Act 1956.

Clause II: The usable quantum of the Narmada water at Sardar Sarovar Dam on the basis of 75% dependability should be assessed at 34,537.44 MCM (28 MAF).

Clause III: Out of the usable quantum of Narmada water (28 MAF), Madhya Pradesh is entitled to a share of 22,511.01 MCM (18.25 MAF), Gujarat to 11,101.32 MCM (9.0 MAF), Rajasthan to a share of 616.74 MCM (0.50 MAF), and Maharashtra to a share of 308.37 MCM (0.25 MAF).

Clause IV: Excess water or distress in any year compared to the utilizable quantum (28 MAF) shall be shared in the same proportion as their actual shares, that is, 73 by Madhya Pradesh, 36 by Gujarat, 1 by Maharashtra, and 2 by Rajasthan.

Clause V: The order is subject to review any time after a period of 45 years from the date of publication of the decision by the tribunal.

Clause VI: The full supply level of Navagam Canal offtaking from Sardar Sarovar should be fixed at 91.44 m (300 ft) at the head regulator with a gradient of 1 in 12,000 up to the offtake of Saurashtra branch, and 1 in 10,000 beyond that point up to Rajasthan border. The bed gradients can be changed by Rajasthan and Gujarat by mutual agreement.

Clause VII: The height of Sardar Sarovar Dam should be fixed at Full Reservoir Level (FRL) 138.68 m (455 ft) with maximum water level at 140.21 m (460 ft).

Clause VIII: Out of the net power produced at Navagam Canal and River Bed Power House (RBPH) on any day, the share of Madhya

Pradesh will be 57%, Maharashtra's will be 27%, and Gujarat's will be 16%.

The capital cost of power portion of Navagam complex (Sardar Sarovar Project) shall be shared by Madhya Pradesh, Maharashtra, and Gujarat in the ratio of 57:27:16.

Clause IX: The requirements of Sardar Sarovar have to be met from the releases by Madhya Pradesh and by inflows from the intermediate catchment, surplus to the requirements of Madhya Pradesh below Narmada Sagar and Maharashtra. The release from Maheshwar works out to 10,015.86 MCM (8.12 MAF). Uniform releases by Madhya Pradesh per month would be 834.65 MCM (0.677 MAF).

Clause X: Madhya Pradesh shall take up and complete the construction of Narmada Sagar Dam with FRL 262.13 m (860 ft) concurrently with or earlier than the construction of Sardar Sarovar Dam.

Gujarat should credit to Madhya Pradesh each year 17.63% of the expenditure on account of the cost of the construction of Narmada Sagar Dam.

Clause XI: This contains detailed directions pertaining to the Sardar Sarovar Project on the definition of oustees, land to be compulsorily acquired, the liability of Gujarat to pay compensation for land acquisition and rehabilitation, provisions for rehabilitation, and programme for payment by Gujarat to Madhya Pradesh and Maharashtra on this account.

Clause XII: The cost of Sardar Sarovar Unit I Dam and appurtenant works should be apportioned between the irrigation and power components in the ratio of 43.9: 56.1.

Clause XIII: The irrigation component cost of Unit I Sardar Sarovar Dam and appurtenant works should be shared by Gujarat and Rajasthan in the ratio of 18:1.

The cost of Navagam Canal with its design approved by the NCA shall be shared between Gujarat and Rajasthan as follows:

- The cost differentiation in respect of land, earth work, and lining due to change in the bed gradient prescribed by the tribunal shall be borne by Rajasthan in full.
- The actual cost differentiation, stated above, shall be shared on cusec mile basis.

Clause XIV: Under this clause, the authority gave detailed directions for setting up of Narmada Control Authority (NCA) and a review committee for securing compliance and implementation of the decision and direction of the NWDT. Under this clause, the tribunal had also advised for setting up of Sardar Sarovar Construction Advisory Committee (SSCAC) to monitor the construction of Unit I Dam and appurtenant works and Unit III power complex of the project.

Sub-clause 17 of Clause XIV reads: 'Nothing contained in this Order shall prevent the alteration, amendment or modification of all or any of the foregoing clauses by agreement between all the States concerned.'

Clause XV: The states of Gujarat, Madhya Pradesh, Maharashtra, and Rajasthan shall bear their own costs for appearing before the tribunal as well as the expenses of the tribunal.

Clause XVI: In addition to the sharing of water (including surplus and distress in any year), the directions by the tribunal relating to the FRL of Sardar Sarovar Dam, the sharing of costs and benefits, regulated releases by Madhya Pradesh for the requirements of Sardar Sarovar Project, payment to be made by Gujarat for regulated releases, and machinery to be set up for the implementation and compliance of its decisions are subject to review at any time after a period of 45 years from the date of the publication of the decisions by the tribunal in the official gazette.

MECHANISM TO IMPLEMENT THE NWDT AWARD

In line with the decision passed in December 1979, the NWDT drew a mechanism under Section 14 for the execution of its directions in the four states. Accordingly, the Ministry of Irrigation issued a notification on 10 September 1980 to constitute the NCA headed by the secretary of the Ministry of Water Resources and Review Committee of the NCA with the Union Minister in charge of water resources. The NCA started functioning with effect from 20 December 1980 and was charged with the responsibilities for monitoring/controlling/executing the following assignments:

- The storage, apportionment, regulation, and control of Narmada.
- Sharing of power benefits from Sardar Sarovar Project.
- Regulated releases by Madhya Pradesh.
- Acquisition of lands and properties likely to be submerged under the Sardar Sarovar Project by the concerned states.
- Compensation, rehabilitation, and resettlement of those affected by this project.
- Sharing of costs.
- The Supreme Court in its judgment dated 18 October 2000 vested the authority in the NCA to give permission for raising the height of Sardar Sarovar Dam beyond 90 m, after obtaining clearances from Environment Subgroup and R&R Subgroup.

SUBGROUPS AND COMMITTEES

A number of subgroups were constituted by the NCA to execute tasks pertaining to different areas. The secretary of the concerned ministry of Government of India heads the particular subgroup. The following subgroups operate under the NCA:

- Environment Subgroup
- Rehabilitation and Resettlement Subgroup
- Water and Metrology Subgroup

In addition to the subgroups, the following committees/subcommittees are also in operation:

- Rehabilitation Committee
- Narmada Main Canal Subcommittee
- Energy Subcommittee
- Environment Subcommittee
- Sardar Sarovar Reservoir Regulation Committee

REVIEW COMMITTEE OF NCA (RCNCA)

A review committee has been constituted under the NCA. This committee can review any decision of the NCA either suo moto, or on the application of any party state, or by the secretary to the Ministry of Environment and Forests. This committee comprises six members, including the chairman, which is given as follows:

- Chairman, Union Minister for Water Resources
- Union Minister, Ministry of Environment and Forests
- Chief Minister, Madhya Pradesh
- Chief Minister, Gujarat

- Chief Minister, Maharashtra
- Chief Minister, Rajasthan

As per the decision taken on 18 October 2000 by the Supreme Court, any issue not decided by the RCNCA shall be referred to the prime minister for the final decision. The NCA publishes an annual report every year about the work done by the four states in consonance with the award and the directions of the NWDT. This annual report is also submitted on the floor of the Lok Sabha and the Rajya Sabha.

PHYSICS AND CHEMISTRY OF NARMADA

The unit used for the measurement of the quantity of water is Million Acre-Feet (MAF). One MAF of water is the quantum of water one foot in height spread uniformly over one million (ten lakh) acres of land. The quantum of water is generally expressed in cubic metre. One MAF of water equals to 1233.48 Million Cubic Metre (MCM). One MCM of water amounts to 10 lakh cubic metres of water. If one were to express one MCM of water in acre feet, it would be 810.71 acre feet. The unit for measurement of the quantum of drinking water is Million Litre Daily (MLD). One MLD of water would mean 10 lakh litres of water per day. One can thus imagine the volume of water available in Narmada, which is assessed to be 28 MAF.

Narmada water is perfectly suitable for irrigation and drinking purposes, although the ill effects of industrialization have been a potential source of threat to rivers. Industrial pollutants leave the waterbodies so polluted that many rivers in North India have either vanished or the water is rendered unsuitable for any human purpose. But amidst all adversaries, it is Narmada itself which is maintaining the quality of its water and making it suitable for human use.

The pH value of Narmada is found to be around 8. The specific conductance is slightly higher and lies in the range of 200–500. The chemical analysis of the water reveals the absence of nitrogen compounds and the desired values of fluoride concentration. This indicates the non-existence of domestic pollution. The lower values of nitrates and phosphates rule out the presence of chemical pollutants in the water. The dissolved oxygen and biochemical oxygen values of Narmada indicate the absence of biological activities. Thus, Narmada, with a little conventional treatment, is suitable for drinking purposes, and comes under A-grade classification. As per the classification for irrigation purpose, Narmada is classified as $C_2 S_1$ type, which is suitable for irrigation.

NARMADA UNDER SEVERE POLLUTION THREAT

Narmada is, at present, somehow maintaining its quality of water despite facing severe pollution threat. The pollution is caused by industrial and domestic pollutants being discharged into the river by industries. A survey revealed that in Madhya Pradesh, where the Narmada flows for a length of 1,077 kms, is receiving 208 million litres of polluted dirty water and more than 529 tons of garbage every day (Urban Administration and Development Department Survey, 2012).* This pollution is being discharged into the Narmada by about fifty-three towns and cities. Besides, industrial waste discharges contribute their fair share in discharge of toxins as well. Thus, Narmada is struggling hard to remain clean. The growing population and the religious activities in the river, uncontrolled mining on the riverbed, and encroachments on the river frontages are some other issues creating trouble for Narmada. As a result, not only the Narmada, but all the tributaries of the river are also reeling from heavy pollution.

Till now, no effective scientific and sustainable approach to this serious issue has been noticed. In October 2014, the central bench of the National Green Tribunal (NGT) asked the government of Madhya Pradesh to state the reasons for failing to control pollution in Narmada.

* Urban Administration and Development Department, Government of MP, conducted the survey in 2012. The survey report was published in the Hindi daily *Dainik Bhaskar* issue, dated 26 June 2012.

The Ministry of Environment, Forest, and Climate Change (MoEF) has been directed to submit details of the progress made in this regard and to send feedback of the central government on the proposals sent by the state governments till date.

The NGT has also asked the government of Madhya Pradesh to file status report on the budget of 12.99 crore that the National River Conservation Directorate (NRCD) of MoEF had sanctioned for the execution of pollution-control projects on Narmada in Hoshangabad in 2007. The NGT has also asked the pollution control board of Madhya Pradesh to file an affidavit with regard to the quality of water in the Narmada prior to and post 2007, and in connection with the work carried out by the pollution control board, as was sanctioned by the central government in 2007. The NGT has ordered the governments of Madhya Pradesh and Chhattisgarh to put a complete ban on the production of polythene made of prohibited microns.

Thus, there is a dire need to develop a scientific and sustainable approach to keep the Narmada healthy for years to come. Experts have opined that a system should be commissioned for the regular monitoring of the biohealth of Narmada in order to take the necessary corrective steps in this direction.

AVAILABILITY OF WATER IN NARMADA BASIN

The NWDT made a plan for Narmada Valley development by assessing Narmada and taking into account 75% dependable flow on the basis of inflows for a period of approximately 80 years (1891–1970). For working out the 75% dependable flow, the NWDT took into account the inflows for the years 1891 to 1947 on the basis of rainfall figures by hindcasting, and for the years 1948 to 1970 on the basis of the observe flow at the gauge and discharge sites at Jamtara, Mortakka, and Garudeshwar. The hindcasting was done as the observe flows were available for the period of only 22 years. In one of the studies in the year 1993, the CWC observed that by looking at the fluctuation in the annual availability of water in the Narmada basin, the correct figure of 75% dependable flow can be worked out by taking into account the data of 130 years.

PROCEDURE OF WATER ACCOUNTING

For regular water accounting, the NWDT has directed that the NCA shall furnish the annual water account for the water year (1 July to 30 June) by the end of August of the next water year. Each party state may make any observation on the account or make corrections, if any, within one month of receipt of the water account. After making the necessary modifications, the authority shall furnish to each party state the final water account for the water year by 31 October. Accordingly, since 1980, the party states furnish the information and the NCA publishes water accounts of each water year. This includes the inflow of water in the Narmada basin, water used by the states for irrigation, municipal, and industrial purposes, and water used by general public and private pumps.

Annual inflows in the Narmada basin till Sardar Sarovar (1980–2016)

Water Year	Total flow (MAF)
1980–81	31.82
1981–82	24.83
1982–83	17.98
1983–84	34.64
1984–85	36.12
1985–86	20.54
1986–87	31.36
1987–88	13.80
1988–89	30.06
1989–90	22.68
1990–91	45.41
1991–92	24.01
1992–93	19.72
1993–94	34.43
1994–95	59.56
1995–96	22.23
1996–97	25.74
1997–98	31.91
1998–99	21.85
1999–2000	42.07
2000–01	13.80
2001–02	23.35
2002–03	22.76
2003–04	33.59
2004–05	18.75

Water Year	Total flow (MAF)
2005-06	37.32
2006-07	35.25
2007-08	30.84
2008-09	19.11
2009-10	21.72
2010-11	23.15
2011-12	35.04
2012-13	41.75
2013-14	61.97
2014-15	27.13
2015-16	23.98
2016–17	33.21

It is clear from the aforementioned table that there are large fluctuations in the yearly inflow. The availability is worked out on the basis of 75% dependable flow. This dependability may be explained by the way that in the given slab of water years, the assessed availability of water would be in 75% water years of the slab, while in the rest 25% water years, there may not be assessed water quantum. Taking into account large fluctuations in the yearly inflow and 75% dependable flow, the sharing of water to the states is determined as follows:

Madhya Pradesh	:	65.19%
Gujarat	:	32.14%
Rajasthan	:	1.78%
Maharashtra	:	0.89%

BIGGEST WATER SHARE TO MP

During proceedings of the NWDT in the year 1972, Madhya Pradesh (MP) had submitted a master plan in which 29 large, 135 medium, and about 3,000 small dams were proposed. Based on this master plan, the NWDT had allocated 18.25 MAF water to Madhya Pradesh out of the assessed 28 MAF water in Narmada basin.

At the time of the final award by the NWDT, the construction of dams mentioned in the master plan rested with the irrigation department of the state. Later, in July 1981, the government constituted Narmada Valley Development Department (NVDA) with a view to take up 29 major projects, exclusively mentioned in the master plan. Later, under the department, Narmada Planning Agency was constituted on 7 December 1982. This agency was merged with NVDA, which was constituted on 16 July 1985. With this, a body higher than the NVDA called the Narmada Control Board (NCB) was constituted. The responsibility of constructing 29 major projects of the Narmada Valley was entrusted to the NVDA. Construction of medium and minor projects was entrusted to the Water Resources Department (WRD). Special administrative and financial powers were delegated to the NCB and the NVDA, as compared to other departments.

In the resolution dated 16 July 1985, whereby the NCB and NVDA were constituted, detailed rules of business were laid down. It was affirmed that the construction of major projects, with a view to exploit the waterbodies and resources in the Narmada Valley, would lie in the jurisdiction of the NVDA. The NVDA would function under the complete authority of the NCB. The broad functions of the NCB entailed supervising the working of the NVDA. This included administrative consent to the working plans, according approval to the delegation of technical and financial powers, and to issue instructions from time to time for efficient conduct of business in the NVDA.

MASTER PLAN OF MADHYA PRADESH

In the master plan submitted by Madhya Pradesh before the NWDT, the classification of projects were done on the basis of their irrigation potential. Those with a potential of irrigation more than 10,000 hectares were classified as major projects, while those with a potential of irrigation less than 10,000 but more than 2,000 hectares were categorized as medium, and those with a potential of irrigation less than 2,000 hectares were the minor projects. The major projects, as mentioned in the master plan, are as follows:

Projects on River Narmada

S. No.	Name of the project	Place of the project/district
1	Rani Avanti Bai Sagar Bargi Diversion	Near village Bargi Jabalpur
2	Indira Sagar	Near village Punasa, Khandwa
3	Punasa Lift	Indira Sagar Canal, Khandwa
4	Omkareshwar	Near Omkareshwar, Khandwa
5	Maheshwar	Near Maheshwar, Khargone
6	Upper Narmada	Dindori
7	Raghavpur	Dindori
8	Basania	Mandla
9	Rosra	Mandla
10	Chinki	Near village Piparha, Narsingpur

Projects on the Tributaries of Narmada

11	Barna	On river Barna, Raisen
12	Sukta	On river Sukta, Khandwa
13	Tawa	On river Tawa, Hoshangabad
14	Kolar	On river Kolar, Sehore
15	Matiyari (Dhobatoria)	On river Matiyari, Mandla
16	Maan	On river Maan, near village Jeerabad, Dhar
17	Shahid Chandra Shekhar Azad (Jobat)	On river Hathni, near village Vaskal, Alirajpur
18	Upper Beda	On river Beda, near village Naimit, Khargone
19	Lower Goi	On river Goi, Barwani
20	Halon	On river Halon, Mandla
21	Sitareva	On river Sitareva, Chhindwara
22	Upper Budhner	On river Budhner, near village Mawye, Mandla
23	Ataria	On river Hiran, near village Ataria, Jabalpur
24	Sher	On river Sher, near village Gangwani, Narsingpur
25	Shakkar	On river Shakkar, near village Hathanpur, Narsingpur
26	Machhrewa	On river Machhrewa, near village Bhadardon, Narsingpur
27	Dudhi	On river Dudhi, near village Dharo Parao, Hoshangabad
28	Morand	On river Morand, near village Moreghat, Hoshangabad
29	Ganjal	On river Ganjal, near village Jawardha, Harda

MEDIUM PROJECTS

A total of 135 medium irrigation projects have also been proposed in the master plan.

MINOR PROJECTS

About 3,000 minor irrigation projects were identified and proposed in the plan.

DIVISION OF LABOUR TO EXECUTE MASTER PLAN

In order to construct proposed irrigation projects and ensure the utilization of allocated water by the year 2024 as per the award, Madhya Pradesh had entrusted the related work to its different departments as follows:

Execution of master plan by different departments

Projects/areas of utilization	**Quantity of water in MAF/MCM to be utilized**	**Department/agency charged with the responsibility**
29 Major Projects (Irrigation)	11.360/14012.33	Narmada Valley Development Authority
135 Medium Projects (Irrigation)	2.880/3552.42	Water Resources Department
3,000 Minor Projects (Irrigation)	2.510/3096.03	Water Resources Department
Industrial	0.743/916.48	Industry Department
Domestic/Municipal	0.757/933.74	Public Health Engineering Department
Total	**18.250/22511**	

UTILIZING NARMADA WATER

To utilize the allocated 18.25 MAF water, Madhya Pradesh, though initially at a slow pace, started the construction of projects on Narmada and its tributaries. The first priority was irrigation followed by hydropower.

PROJECTS ON NARMADA RIVER

Rani Avanti Bai Sagar Project (Irrigation and Hydroelectric)

Rani Avanti Bai Sagar Dam has been constructed on Narmada near village Bargi, district Jabalpur. The dam construction was started in 1971 and completed between 1988 and 1989. The height of the main masonry/concrete dam is 69.50 metres. The main left-bank canal is 135.50 kms long. There are four branch canals from the main canal. It is estimated to create an irrigation potential of 1.57 lakh hectares of land spread over the districts of Jabalpur and Narsingpur.

Bargi Diversion Project (Irrigation)

The right-bank canal project of Rani Avanti Bai Sagar Dam is known as Bargi Diversion Project. The right-bank canal would be 196.8 kms long. This canal project is an inter-basin project, which takes the Narmada water from Narmada basin to Rewa and Satna districts, whose command areas fall under the Ganges basin. Bargi Diversion would create an irrigation potential of 2.45 lakh hectares in the districts of Jabalpur, Katni, Rewa, and Satna. Besides the main canal, 183.37 kms of branch canals and 2,700 km of minors' distribution network would be constructed. As of now, 104 kms of the main canal is constructed and a 12 km-long irrigation tunnel is under construction. This tunnel would facilitate inter-basin transfer of water, that is, from Narmada basin to Son basin. The construction of the distribution network is underway.

Madhya Pradesh has been urging the central government that the Bargi Diversion Project, being an inter-basin one, should be declared a national project. However, the initiatives taken for this did not result in the desired outcome. The project is unlikely to be notified as a national project as per information pouring down from the Ministry of Water Resources. For a national project, 90% of the total cost of project has to be borne by the central government.

Indira Sagar Project (Irrigation and Hydroelectric)

Indira Sagar Dam has been constructed on Narmada near village Punasa in Khandwa district. The dam reservoir is the largest man-made reservoir of the country and the third largest reservoir in Asia with an area of 913.48 sq. kms. The dam has a water-holding capacity of 7.9 MAF (12.225 Billion Cubic Metre). It is the mother dam for all other major dams. The foundation of the project dam was laid on 23 October 1984 by then Prime Minister Indira Gandhi.

The construction of the dam could be completed only in the year 2005. The length of the main concrete dam is 840 metres with a maximum height of 92 metres. The dam has the provision of only right-bank canal, measuring up to 248.65 kms in length. The main canal with its distribution network would create an irrigation potential of 1.23 lakh hectares. Construction of the main canal along with its distribution network is being undertaken in four phases, which is mentioned as follows:

A view of Indira Sagar Project Dam and powerhouse water outlet

The four phases of the construction of Indira Sagar Project

Phase	Length of the main canal	Irrigation potential
Phase I	0 to 71 kms	36,100 hectares
Phase II	71 to 142 kms	26,100 hectares
Phase III	142 to 206 kms	20,700 hectares
Phase IV	206 to 248.65 kms	40,100 hectares

As of now, the construction of 160 kms has been completed. Efforts are ongoing for the completion of the main canal and its distribution network. Upon the completion, about 69,000 hectares of land in Khargone district, 36,000 hectares of land in Barwani district, and 19,000 hectares of land in Khandwa district would be irrigated. Keeping this in view, provision has been made for keeping 0.06 MAF of water from the reservoir for domestic and industrial purposes.

Khargone Lift Irrigation

For the total irrigation potential of 1.23 lakh hectares of Indira Sagar Project, the net contribution of 33,140 hectares would be made by Khargone Lift. Under the scheme, water would be lifted from the main canal of Indira Sagar at 92.10 kms through pumps upto a height of 38.62 metres. The water will then pass through a 83 km long lift canal. The head discharge would be 20.50 cubic metres per second.

Punasa Lift Irrigation

Punasa Lift Irrigation is the replacement of a project on river Tawa, which was proposed in the master plan. Under the original project, water was to be lifted from the proposed dam on Chhoti Tawa River and irrigate the highlands near the head regulator of Indira Sagar Canal. After examining the proposal, it was found that instead of constructing a small dam on Chhoti Tawa and lifting water from there, the better alternative would be to lift water from Indira Sagar Canal. This project was named the Punasa Lift Irrigation Project. Under the scheme, a control structure and a jack well were built on the head of the main canal of Indira Sagar near village Chandel in Khandwa district. Water would be lifted there in three stages and fed to four balancing reservoirs.

A total of 780 km distribution network would be made by means of underground pipes. The project would create an irrigation potential of 35,000 hectares of land. Power to run this lift irrigation scheme would be made available from the 15 megawatt canal bed powerhouse on Indira Sagar main canal. As of now, the work of Punasa Lift Project is 98% completed.

Omkareshwar Project (Irrigation and Hydroelectric)

Omkareshwar Dam is situated on Narmada at the famous pilgrimage centre Omkareshwar in Khandwa district. The main concrete dam is 53 metres high and 949 metres long. There is a common carrier canal (12.39 kms long) from the left bank of the reservoir. The common carrier canal would have the 162.35 km long canal on the right side and 64.11 km long canal on the left. The common carrier would create an irrigation potential of 960 hectares, left-bank canal 20,580 hectares, and the right-bank canal 70,630 hectares. Thus, a total of 92,170 hectares irrigation potential would be created through the carrier canal and the right- and left-bank canals.

Omkareshwar Lift

There is a provision for the construction of 125 km long-lift canal which would run parallel to the right-bank canal. The lift canal would create an irrigation potential of 58,020 hectares. Water would be lifted from the right main canal extending up to a length of 97.75 kms, for which a junction structure is constructed. At a height of 56 metres and at a pace of 15 cubic metres per second, water would be lifted. For this purpose, five motor pumps, each of 1,200 horsepower (hp) have been being installed. Approximately, 11.20 megawatt of power would be consumed in lifting the water. The water so lifted would be fed to Sisliya Tank (balancing reservoir) about 10.20 kms away. From there, water would be discharged for gravity flow in to the lift canal.

The entire canal system of Omkareshwar Project, including the lift, would create an irrigation potential of 1,46,800 hectares. The district-wise command area would be 73,600 hectares in Khargone, 73,000 hectares in Dhar, and 200 hectares in Khandwa district. Besides, there is a provision for diverting 74 lakh cubic metres of water from Omkareshwar reservoir for industrial and domestic uses. The foundation stone of Omkareshwar Project was laid on 30 August 2003 by then Prime Minister Atal Bihari Vajpayee.

Maheshwar Project (Hydroelectric)

The Maheshwar Project is purely a hydroelectric project with no irrigation provision. It is last in the series of major dam projects on Narmada in Madhya Pradesh. Under this, a concrete dam of 3,420 metres length and 36 metres height has been constructed on Narmada near Maheshwar in Khargone district. The project would generate a capacity of 400 megawatt of power. From Maheshwar Dam, 8.12 MAF water would be released against 9 MAF water allocated to Gujarat, specified in the NWDT award. The remaining quantity of water would be supplemented by the catchment area falling within Gujarat itself.

Omkareshwar Project Dam on River Narmada

In the beginning, the Maheshwar Project was to be constructed by the NVDA. Later, in December 1988, the construction of the Maheshwar Project was entrusted to the energy department. The energy department in turn assigned the work of the project to Maheshwar Hydel Power Corporation Ltd. The company was owned by S. Kumars. The construction of the project was completed in December 2012, but power generation was withheld because the NBA had objected on the grounds that rehabilitation of the affected families had not been done. The NGT, after hearing the petition of the displaced people affected by the Maheshwar Dam on April 2014, had directed the company not to close the gates of the dam. The company admitted that the lake of resources became a hindrance in its rehabilitation work.

Upper Narmada Project (Irrigation)

The Upper Narmada Project is proposed to be on Narmada in the village of Reenatola, Dindori district. The dam site is about 72 kms away from Amarkantak, the place where Narmada originates. The proposed concrete dam would be 33.8 metres high with a length of 1,776 metres. Two canals, 86.50 km long left bank and 65.5

km long right bank, would be constructed from the reservoir. The project would create an irrigation potential of 18,616 hectares. The irrigation benefit would be available to the districts of Dindori and Anuppur.

Raghavpur Project (Irrigation and Hydroelectric)

The Raghavpur Project is planned to be constucted on Narmada in Mandla district at a site of 93 kms from the district headquarters, 14 kms away from the Mandla–Dindori state highway. A detailed project report is underway.

Basania Project (Irrigation and Hydroelectric)

The site for the proposed Basania Project is located on Narmada in Mandla district at a distance of 50 kms from the district headquarters, 15 kms away from the Mandla–Dindori state highway. The preparation of a detailed project report is underway.

Rosra Project (Irrigation and Hydroelectric)

The Rosra Project, to be constructed in Mandla district on Narmada, is also at the stage of preparation of a detailed project report.

Chinki Project (Irrigation)

Under this project, a large dam was proposed to be built, submerging 5,937 hectares of farmland, creating an irrigation potential of 73,979 hectares in Narsingpur and Raisen districts. The submergence was highly opposed by the local farming community. Therefore, in 2016, it was decided not to build the dam. It was concluded that water would be directly lifted from the powerful flow of the river at the site. The lift would create an irrigation potential of 1.04 lakh hectares.

PROJECTS ON NARMADA'S TRIBUTARIES

Barna Project (Irrigation)

The Barna Irrigation Project is constructed in village Badi on river Barna. The work of this project was started in December 1968 and completed in June 1997. Under the project, a masonry and a concrete dam with a height of 47.70 metres was constructed. The length of the left-bank canal and the right-bank canal of Barna Dam is 29 kms and 10.15 kms respectively. The project had created an irrigation potential of 60,750 hectares in the districts of Raisen and Sehore.

Sukta Project (Irrigation)

The Sukta Irrigation Project is built on river Sukta in village Khirala of Khandwa district. The construction of the project was started in April 1973 and completed in March 1985. Under the project, a masonry-cum-earthen dam with a height of 32 metres and a length of 22.9 metres was constructed. The spillway exists on the right side. On the left side, there is a 26 km long canal. The project had created an irrigation potential of 16,599 hectares in Khandwa district.

Tawa Project (Irrigation and Hydroelectric)

The Tawa Irrigation Project is constructed on river Tawa in the district of Hoshangabad. The construction of the project was started in 1956 and completed in 1998. The left-bank canal has a length of 131 kms and the right bank 7.5 kms. Besides, there are branch canals to Piparia and Baghera of length 60 kms and 24 kms respectively. The Tawa Project had created an irrigation potential of 2,42,915 hectares in Hoshangabad and Harda.

Kolar Project (Irrigation)

The Kolar Irrigation Project was constructed on river Kolar in village Lavkhandi of Sehore district. The construction of the project was started in 1979 and the work of the main dam was completed in June 1991. The masonry dam is 45 metres high and 1,242 metres long. The length of the left-bank canal and the right-bank canal are 29.5 kms and 24.7 kms respectively. Besides irrigating 33,593 hectares in Sehore, the project provides drinking water to the city of Bhopal.

Matiyari Project (Irrigation)

This project came in place of the Dhobatoria Project originally proposed in the master plan. Matiyari Dam was constructed on river Matiyari in the village of Simaria in Mandla district. The dam work was started in 1978 and completed in 1991. The main dam is earthen, whereas there is concrete spillway on the right side. The right-bank canal of the dam is 37.71 km long, while the length of the left bank is 35.76 kms. This project had created an irrigation potential of 10,120 hectares in Mandla district.

Maan Project (Irrigation)

The Maan Project is constructed on river Maan, a tributary of Narmada in village Jirabad. The project includes masonry-cum-earthen dams. The length and height of the main masonry dam are 138 metres and 53 metres respectively. While the length of the left-bank canal is 16.27 kms, the right bank canal is 10.92 kms. The construction of the project dam was started in the year 1989 and completed in 2006. It had created a potential of irrigating 15,000 hectares in the district of Dhar.

Shaheed Chandra Shekhar Azad, Jobat Project (Irrigation)

The Jobat Project was constructed on river Hathni in the village of Vaskal, Alirajpur district. The length of the main masonry dam is 148 metres and its height 38.6 metres. On the left side of the main dam is an earthen dam of length 169.5 metres and that of the right side is 60 metres. The main canal, measuring 29.73 km in length and constructed on the left side of the dam, creates an irrigation potential of 9,848 hectares of land in the tehsil Kukshi, Dhar district. The construction of the project was started in 1997 and completed in the year 2008.

Upper Beda Project (Irrigation)

The Upper Beda Project is constructed on river Beda in village Nemit, which is at a distance of 17 kms from Bhikangaon tehsil headquarters of Khargone district. The project comprises 191 metre long and 35 metre high main masonry, a concrete dam, and 2,223 metre long earthen dam. It is proposed to construct a long left-bank canal of 30.427 kms with a network of 22 km long-branch canals and sub-canals of a total length of 80 kms. The work of the main canal and network is ongoing. This canal system would create an irrigation potential of 9,917 hectares.

There is a provision for irrigation on the right side too. It has been proposed to lift water on the right side by means of 16 mm diameter pipes over a distance of 170 metres and then to release it in the 8 km long, open right-bank canal. This canal, with the network of sub-canals and distribution branches, would create an irrigation potential of 1,500 hectares.

Lower Goi Project (Irrigation)

The Lower Goi Project has been proposed on river Goi near village Panchpola, 21 kms away from the district headquarters of Barwani. Under the project, a long concrete dam of 203.50 metres and a long earthen dam of 2226.50 metres would be constructed.

The height of the concrete dam would be 43.80 metres. From the right side of the dam, water shall be taken through a 5.70 km long tunnel to the 30 km long main canal, and, thereafter, the distribution network of the sub-canals running for a total length of 32 kms would irrigate 13,760 hectares of land. The command area is spread over 51 villages in the tehsils of Rajpur, Anjad, and Barwani coming under

Barwani district. The construction of the dam had started in February 2009. As of December 2013, 60% of the dam has been completed.

Halon Project (Irrigation)

The Halon Project is proposed to be built on river Halon in village Karanjiya of Bichhiya block in Mandla district. The proposed main masonry and concrete dam would be 100 metres long and 31 metres high. The left-bank main canal, measuring 81.50 km in length, is proposed to be constructed from the reservoir. The project would create an irrigation potential of 13,040 hectares in Mandla. As of December 2013, the initial work of the dam has started.

Ataria Project

The Ataria Project is proposed to be constructed on river Hiran in district Jabalpur. The project site is located at a distance of about 1.25 km from Sihora Railway Station, which falls on the Mumbai–Howrah railway line. Earlier, in order to achieve the proposed irrigation potential of 5,900 hectares, the height of the dam was decided to be 407 metres. The intended height of the dam was going to submerge three villages, completely immersing 3,587 hectares of agricultural land. This submergence created strong opposition from the farmers. In such a situation, the project was reviewed in 2009 and, thereafter, a decision was reached to reduce the height of the dam by 4 metres. As a result of the reduction of the dam height, no village would fully get submerged and the total submergence would get reduced to only 1,028 hectares. With the change, the irrigation potential also got reduced to only 1,986 hectares.

In order to augment the irrigation potential, a feeder canal was proposed to be constructed from the reservoir. The feeder canal would supply water to the right-bank of the main canal (Bargi Diversion) of Rani Avanti Bai Sagar Project at a chainage of 81 kms. This additional quantity of water will extend the area of irrigation by the Bargi right-bank canal by about 9,000 hectares.

OTHER PROPOSED PROJECTS

The master plan also includes the following projects that are yet to be taken up for construction:

1. Sitarewa Project in district Chhindwara
2. Upper Budhner Project in district Mandla
3. Sher and Machhrewa projects in district Narsingpur
4. Shakkar Project in district Chhindwara
5. Dudhi Project in district Hoshangabad
6. Morand Project in district Hoshangabad
7. Ganjal Project in district Harda

MEDIUM PROJECTS

Out of 135 medium irrigation projects proposed in the master plan, the following projects have been completed as of December 2013:

Medium irrigation projects completed as of December 2013

S. No.	Name of project	District
1	Jalgaon	Balaghat
2	Banjar	Balaghat
3	Bichhiya	Mandla
4	Dhuandhar	Mandla
5	Bahoriband	Katni
6	Mehgaon Tola	Jabalpur
7	Barnu	Jabalpur
8	Madai	Jabalpur
9	Pariyat	Jabalpur
10	Dukrikheda	Hoshangabad
11	Chandrakeshar	Dewas
12	Paras	Dewas
13	Sampna	Betul

S. No.	Name of project	District
14	Choral Nakhedi	Indore
15	Segwal	Barwani
16	Satak	Khargone
17	Dejla-Dewda	Khargone
18	Kunda	Dhar
19	Sakalda	Dhar

MINOR PROJECTS

Out of 3,000 minor projects proposed in the master plan, 1,032 projects were completed till December 2013.

LIFTING NARMADA WATER TO MALWA

The region of Malwa in Madhya Pradesh has been facing severe water shortage. The plummeting groundwater has dried up the traditional water sources. The once perennial rivers of the region have been shrinking and converted into *nullah*s. Experts have opined that if such conditions prevail, Malwa may come under the clutches of famine-like conditions in years to come.

The people of Malwa have always looked towards Narmada as the only solution to their plight. For many years, they have been demanding to bring Narmada water to Malwa. The demand was considered impractical and difficult, as Malwa plateau is at a higher elevation from where Narmada flows. In the decade of the 1990s, it was proposed to lift 3 MAF water from Indira Sagar reservoir to a height of 300 metres and to release it at an appropriate place in the Malwa plateau. The idea was to raise the groundwater table and recharge river Kshipra.

The proposal had been under technical and financial scrutiny for a long time. Discussions and high-level meetings were held at various levels. Initially, the project was rated as quite expensive. It was estimated that lifting 3 MAF water to a height of 300 metres would consume about 552.72 megawatt of power. The cost of power and maintenance of the project would be about Rs. 1,200 crore per year. After extensive exercises on cost and benefit, it was concluded that the proposal was extremely expensive and impractical.

Although the proposal of the project was rejected by the then government, the people of Malwa never abandoned the demand of transferring Narmada water to Malwa. With each year, the diminishing level of groundwater has given rise to scarcity of water in the entire Malwa region. The mounting pressure of the people drove public representatives, politicians, and administrators to find a permanent solution to the problem.

In the year 2012, an Indian Administrative Service (IAS) officer Rajneesh Vaish took charge of the NVDA as the vice chairman (chief executive officer). The officer, having strong technical background, thought over the issue and came out with an idea which was worth implementing. The idea was soon converted into a Detailed Project Report (DPR). The feasibility of the plan costing less than the initially thought project to lift 3 MAF of water made it highly practicable. The project was named Narmada-Kshipra-Simhastha Link, since Narmada water would be powerlifted to Kshipra River, and through Kshipra, water would be available for holy baths during religious congregations at Ujjain during Simhastha Mela. It was also proposed that in later stages, water would be lifted for microinjection in the river basins of Parvati, Kali Sindh, and Gambhir of Malwa region.

NARMADA-KSHIPRA-SIMHASTHA (NKS) LINK

With the benefits of this project available to a large population, the Malwa state government accorded administrative and financial sanctions for the project on top priority.

To see Narmada water flowing in the rivers of Malwa had been the dream of the residents. This approach of the government to

The eye-catching point of Narmada–Kshipra confluence near Indore in MP

bring Narmada water to Malwa plateau, which would then flow into river Kshipra, was highly appreciated by the people. At last, after 14 months, this dream project was completed and successfully tested on 6 February 2014.

Under the project, 5 cumecs of Narmada water is powerlifted from Sislia tank under Omkareshwar Irrigation Project. Water is being lifted to the height of 348 metres and transported through mild steel pipes to the origin of Kshipra River at a distance of about 48 kms. Three power stations, two each of 9,000 and one of 1,030 kW, are being commissioned. The required power would be taken from 15 megawatt Canal Head Power House (CHPH) of Indira Sagar Project. The water is being taken to the origin of river Kshipra in the Vindhyachal Mountains. There are two exit structures: one discharges 4.8 cumecs of water in to the river, while the other discharges 0.2 cumecs of water. This water is taken to flow through artistically constructed waterfalls, creating scenic beauty. The water so flown through the falls is then passed through a spiral-shaped canal in traditional river ghats. This water also confluences with river Kshipra.

The river Kshipra, boosted by river Narmada, would provide drinking water to more than 250 villages on the banks of the river, along with drinking water to the cities of Ujjain and Dewas. Water would also be made available to the industries of this region, including the industries of Pithampur, which is a big industrial centre of Malwa. The cost of the project amounts to Rs. 432 crore and is designed to fulfil the drinking and industrial needs of the area for years to come.

HYDROPOWER

Hydroelectricity generation is an important benefit. The generation capacity of hydroelectricity from the entire Narmada basin (including Gujarat and Madhya Pradesh) has been assessed to be 3846.4 megawatt.

To exploit the major sites of Indira Sagar and Omkareshwar, Madhya Pradesh decided to use the expertise of National Hydroelectric Power Corporation (NHPC). For this, a joint venture was formed under the name of Narmada Hydroelectric Development Corporation Ltd (NHDC). NHDC came in to existence in August 2000.

The capital of the NHDC is shared to the extent of 51% by NHPC and 49% by the government of Madhya Pradesh. The NHDC was entrusted with the work of constructing Indira Sagar and Omkareshwar hydroelectricity projects. The NHDC has completed the construction of Indira Sagar Dam in the year 2005, a year before the time prescribed, and Omkareshwar Project Dam in the year 2007. The government of Madhya Pradesh had entrusted the work of 1,320 megawatt thermal powerhouse to this company in the periphery of Indira Sagar reservoir in Khandwa district. The entire electricity generated by the proposed thermal powerhouse would be available to Madhya Pradesh.

NHDC has also identified a site in Kukroo village in Betul district of Madhya Pradesh for the construction of a 100 megawatt wind powerhouse.

The hydroelectricity projects, completed and proposed in Madhya Pradesh, are as follows:

Rani Avanti Bai Sagar

This project has been producing hydroelectricity since 1989. The installed capacity of the main powerhouse is 90 megawatt, comprising two units, each of 45 megawatt.

The canal's head powerhouse on the left bank of the dam, comprising two units each of 5 megawatt each, has the installed capacity of 10 megawatt and is producing power since 2006. Thus, the total installed capacity of the project is 100 megawatt.

Indira Sagar

The work of the Indira Sagar Dam and powerhouse was entrusted to the NHDC on 25 November 2000. The production of hydroelectricity from the installed two units was inaugurated on 18 January 2004 by the then chief minister Uma Bharti. All the eight units of the project, each with a generation capacity of 125 megawatt, were installed in January 2005. Thus, the total installed capacity for the production of power is 1,000 megawatt. The construction of the powerhouse on the canal of Indira Sagar is also completed. The powerhouse has a total capacity of 15 megawatt from three units, each with a capacity of 5 megawatt.

Omkareshwar

The construction of the dam and the powerhouse of Omkareshwar hydroelectricity project were also done by NHDC. The dam and the powerhouse were completed by the NHDC in November 2007. Eight units, each with a capacity of 65 megawatt, have been installed. Thus, the total installed capacity of the project is 520 megawatt. A powerhouse of 5 megawatt capacity is proposed on the canal.

Maheshwar

Maheshwar hydroelectricity project is the last project in the line of major projects to be constructed on Narmada in Madhya Pradesh.

According to NWDT award, 8.12 MAF of water has to be released for Gujarat in a regulated way by Madhya Pradesh from Maheshwar Project. In December 1988, the construction of the Maheshwar Dam was transferred from NVDA to the state department of energy. The energy department entrusted the work of the project in 1992 to Maheshwar Hydel Power Corporation Ltd, owned by S. Kumars. The dam and the powerhouse had been completed by the agency. The Maheshwar Project would add a capacity of 400 megawatt to the state power potential. Since this is a run of the river project, there would not be a large reservoir.

Tawa

Tawa is a tributary of Narmada. An irrigation dam has been constructed on Tawa River. A powerhouse with the capacity of 13.5 megawatt has been constructed on the canal of Tawa Project in Hoshangabad district.

The following projects listed have been also proposed to generate hydropower:

Raghavpur Project	18 megawatt
Rosra Project	20 megawatt
Basania Project	16 megawatt
Chinki Project	25 megawatt
Sitarewa Project	15 megawatt
Gopalpur Project	25 megawatt

Besides the given list, six other sites have been identified. It is proposed to generate 34 megawatt of power collectively out of these potential sites.

In order to explore the remaining possibilities of hydroelectricity in the Narmada basin, the government of Madhya Pradesh got a study done by the centre for hydroelectricity at IIT Roorkee in 2009. This study identified 182 new small and medium hydroelectricity project sites. It was assessed that these freshly identified sites would yield nearly 186.11 megawatt of hydroelectricity.

THE NWDT AWARD FOR GUJARAT

The NWDT, in its final award, allocated 9 MAF water to Gujarat, and in order to avail it directed that the height of the Navagam Dam be kept at 455 ft (138.68 metres). After the award, Gujarat took quick steps in this direction. In the year 1979, immediately after the award, Gujarat constituted Narmada Planning Group (NPG) under the Narmada Development Department. The purpose of constituting NPG was to prepare a well-considered, far-reaching, and comprehensive plan for Sardar Sarovar Dam. It was a plan to offer momentum to the development of Gujarat by way of irrigation and hydroelectricity.

NARMADA PLANNING GROUP (NPG)

In NPG, services of experts in fields like irrigation engineering, geology, environment, forestry, ecology, economies, statistics, public policy, etc. were availed as members of the group. The NPG took help of subject experts, advisors, universities, and technical institutions. The NPG got 25 extensive and in-depth technical studies done. Besides this, 20 socio-economic studies were conducted. Systematic analysis was done in the context of the project, encompassing conclusions of research and studies made earlier both at the level of Government of India and the state government. Every aspect of the project was taken into consideration while preparing the plan. Eventually, within three years of its constitution, the NPG finalized the Sardar Sarovar Development Plan (SSDP) in May 1983.

The SSDP was historic and important in the sense that its planners had incorporated innovations in the traditional mode of thought and made new initiatives in design, construction, operation, maintenance, and other aspects in the planning of the dam projects. Thus, this plan was far superior to the development plans of other dams made earlier. The World Bank also appreciated the SSDP. The planners and experts who made remarkable contribution while being members of the NPG are given as follows:

Planners and experts of Narmada Planning Group

Dr Y.K. Alagh	Executive Vice Chairman (Planning and Economics)
Dr S.R. Hashim	Member (Economics)
Dr B.N. Vishwanath	Member, part-time (Environment and Regional/Hinterland Plan)
Dr S.S. Mehta	Member, part-time (Economics)
Dr N.R. Patel	Member, part-time (System)
J.P. Nayegamegawattala	Advisor (Irrigation Engineering)
P.B. Buch	Advisor (Statistics)
Dr G.A. Patel	Advisor (Agriculture)

SARDAR SAROVAR PROJECT ADVISORY COMMITTEE (SSPAC)

For the appropriate and qualitative construction of the project, the NWDT in its award had given directions for the constitution of an advisory committee for Sardar Sarovar Project. Accordingly, Government of India constituted the Sardar Sarovar Project Advisory Committee (SSPAC) in 1980. The secretary to the Ministry of Water Resources was made the head of the committee. Apart from the representatives of the government and the NCA, the committee comprises of representatives from the state governments of Gujarat, Madhya Pradesh, Maharashtra, and Rajasthan. The secretariat of the committee is located at Vadodara. There is a full-time secretary of the rank of chief engineer and is also the member of the committee. The responsibility of the SSPAC is to ensure the quality and economy in the construction of Unit-I and Unit-III dams and the generation of power as well as extending cooperation in timely completion thereof.

SARDAR SAROVAR NARMADA NIGAM LTD

In order to ensure the construction and the quality of the various works of Sardar Sarovar Project, the government of Gujarat, on 24 March 1988, constituted the Sardar Sarovar Narmada Nigam Ltd under the provisions of the Companies Act 1956. The company comprises of 11 directors, besides the chairman and the vice chairman. They include director (Finance), director (Canal), and director (Civil). There is a managing director and secretary of the company. It is headquartered at Gandhi Nagar. The company, for the conduct of its business, also acquires share capital. The authorized share capital of the company is Rs. 12,500 crores.

Sardar Sarovar Dam on River Narmada in Gujarat

NAVAGAM DAM PROJECT (SARDAR SAROVAR PROJECT)

The then prime minister of India, Pandit Jawaharlal Nehru, had laid the foundation of this dam project on 5 April 1961. But the work on the Navagam Dam could not be started till 1987. Later, the project was named after Sardar Vallabhbhai Patel, the first home minister of the country who hailed from Gujarat.

The dam site of the project is located in tehsil Nanded of Narmada in Gujarat. The project includes 1210.02 metre long and 163 metre high (above the lowest foundation) *pakka* dam, canals, and water distribution system, riverbed powerhouse, canal head powerhouse, and other related structures. The crest level (gates are installed above this level) of the main dam is 121.92 metres. After installation of the gates (the NCA accorded permission to install the gates on 12 June 2014), the FRL of the dam would be 138.68 metres. Total, 30 radial gates would be installed on the main dam. While seven of these gates would be 60 ft x 60 ft in size, the size of the remaining 23 gates would be 60 ft x 55 ft. As much as 61 million cubic metre of concrete has gone into the making of the main dam, consuming 300 crore bags of cement and 5,000 metric tons of steel. The Sardar Sarovar reservoir is 214 km in length and 16.1 km wide. The water storage capacity of the reservoir is 4.73 MAF.

CANAL SYSTEM

The huge canal network of the project comprises 458.318 km long main canal, 2,500 km long 44 branch canals, 5,500 km long distributaries, and about 30,000 km long minors and sub-minors. Besides, there is a huge network of field channels. The total length of Sardar Sarovar canal network is about 85,835 kms, which is larger than the 62,000 km long railway network of India at present.

Under the canal system, two tunnels (bypass tunnels) have been constructed through the hillock on the right side of the reservoir. The diameter of the two tunnels is 5.5 metres with a water flow capacity of

10,000 cubic ft per second. The water passing through these tunnels is released in the first irrigation tank. The link canal takes water from the first irrigation tank to the second. Again, through another link canal, water is tranferred from the second irrigation tank to the third, and yet another link canal takes it from the third to the fourth. Finally, water is released in the main canal from the fourth irrigation tank. 'Godbole waste weir' has been constructed on the third irrigation tank to pump out the extra water by an automatic exit system.

The main canal of the Sardar Sarovar Project is one of the longest lined canals in the world. This 458.318 km long canal, which goes to the border of Rajasthan, was constructed in two phases. The first phase ran from 0 to 144.50 kms and the second from 144.50 to 458.318 kms. The construction of the main canal has been completed. The entire canal has been lined with concrete cement and is properly streamlined for easy water flow. In the Narmada Tribunal Award, there is a provision for releasing the allocated 0.50 MAF water to Rajasthan from the tail end of the canal on the border of Rajasthan. The irrigation potential of Sardar Sarovar Project is 21.08 lakh hectares. Out of this, 18.62 lakh hectares of land in Gujarat and 2.46 lakh hectares of land in Rajasthan shall be irrigated. Madhya Pradesh and Maharashtra would not get irrigation benefits from this project.

Gujarat has started providing Narmada water to the rural and urban populace for domestic use. So far, drinking water facility through the Narmada canal has been provided in 9,633 villages of 16 districts and 131 urban centres, including those in Jamnagar, Anjar, Bhavnagar, and Talaja.

HYDROELECTRICITY

In addition to the 21.08 lakh hectare irrigation potential, Sardar Sarovar Project has hydroelectricity generation capacity of 1,450 megawatt.

As per the NWDT award, the Sardar Sarovar Hydroelectricity Project has been built jointly by Gujarat, Madhya Pradesh, and Maharashtra. The power generation of the project has been shared to the extent of 57% by Madhya Pradesh, 16% by Gujarat, and 27% by Maharashtra.

The project comprises of Sardar Sarovar main dam and an underground riverbed powerhouse. The powerhouse, comprising six Francis turbines, each with a capacity of 200 megawatt, has the total installed capacity of 1,200 megawatt. The turbines are designed to operate in a reversible way too, so as to use water again for power generation. For the reuse of water, a barrage has been constructed downstream at a place called Garudeshwar to collect water coming out of the powerhouse. On running the turbines in the reverse mode, the water collected in the barrage is brought back to the dam. Thus, it may be reused for power generation. During times of acute need of power, this arrangement of using the same water repeatedly produces additional power.

A powerhouse has also been constructed on the main canal of the dam, which has five couplon-type turbines, each with a capacity of 50 megawatt (5x50), making total installed capacity of 250 megawatt. Thus, the total installed capacity of Sardar Sarovar Hydroelectricity Project is 1,450 megawatt.

For the transmission of power, a switchyard complex and a network of transmission lines have been commissioned. The power generated by the project would be distributed in the same ratio in which the three states will bear the expenditure incurred for the project. That is to say, from the power generated through the project, the proportion for Madhya Pradesh will be 57%, Gujarat 16%, and Maharashtra 27%.

For the purpose of discharge of power from Sardar Sarovar Complex, two double lines of 400 kilovolt (kV) have been laid for the transmission of power to Madhya Pradesh, Maharashtra, and Gujarat. The running and maintenance of the complex is done by Gujarat State Electricity Company Ltd (GSECL). A memorandum of understanding has been signed between Sardar Sarovar Narmada Nigam Ltd. and Gujarat State Electricity Corporation Ltd. for this purpose.

SUPERVISION OF ELECTRICITY SHARING

The NWDT entrusted the responsibility to the NCA for issuing guidelines for the generation of power from Sardar Sarovar Project, commissioning of transmission lines and system, and providing electricity to Madhya Pradesh, Maharashtra, and Gujarat. To this end, on 15 June 1983, the NCA constituted a power subcommittee under the chairmanship of its member (Department of Energy). This subcommittee comprised representatives of beneficiary states, Central Electricity Authority (CEA), Western Regional Power Committee, concerned state electricity boards, and Department of Energy. The power subcommittee supervises the related works and issues necessary directions.

ENERGY MANAGEMENT

In order to execute works pertaining to various aspects connected with the generation and distribution of power from Sardar Sarovar Project, the NCA has established an Energy Management Centre (EMC) in Indore. This centre is being run in coordination with Western Regional Power Committee, Western Regional Load Dispatch Centre, Mumbai, State Load Despatch Centre, and other related institutions of the beneficiary states. A subdivisional wing of the EMC has been established at Kevadiya in Gujarat. The management units, thus established at Indore and Kevadiya, besides coordinating in the generation of power from the project and transmission thereof on a daily basis, supervises the distribution of power amongst the three states in the ratio determined by the NWDT in its award.

RAJASTHAN

The government of Rajasthan has constructed a 74 km long Narmada canal in continuation to the 458.138 km long Narmada canal in Gujarat. In the territory of Rajasthan, the Narmada canal starts from village Silu in tehsil Sanchore of Jalore district. The earthwork, lining, and structures in the entire 74 km long canal has been completed, including the Luni Canal Siphon. The NWDT had allocated 0.50 MAF Narmada water to Rajasthan. This quantity of allocated water shall be made available to Rajasthan at its border by Gujarat through the Narmada canal.

Initially, the allocated water was proposed for irrigating 73,157 hectares of land in the drought-affected districts of Jalore and Barmer. Now the target of irrigation has been revised by Rajasthan. The revised target is aimed to irrigate 2.46 lakh hectares of agricultural land. To achieve this target, nine distributaries for flow irrigation and three distributaries for lift irrigation are taking off from the main canal. Work on the distribution network is in progress.

Gujarat had started supplying water to Rajasthan since 27 March 2008. The entry of Narmada water into the territory of Rajasthan was ceremonial. Narmada water was welcomed by the chief minister of Rajasthan, Smt. Vasundhara Raje, with religious rituals. The release of water into Rajasthan is being controlled by a cross regulator built at a chainage of 424.34 kms on the main canal. Water is released into Rajasthan at the rate of about 500 cubic ft per second. Under the Narmada canal project, there is no provision for building watercourses in order to channel water to the fields. Irrigation is done by sprinkler method. A chamber and pump house have been constructed on every chunk of land. Through the pipes, water is pumped to the fields under pressure. The cultivators have established the sprinkler system at their own expense to take the benefit of irrigation.

As the main canal and the distribution system pass through sandy regions, there is an acute problem of the seepage of water from the canals. Because of this, the entire canal system is being lined with concrete. Besides providing the benefit of irrigation from Narmada canal, 0.1064 MAF water is also allocated for drinking purpose to 1,336 villages in the districts of Jalore and Barmer and 3 cities in Jalore, thereby covering 14 lakh population.

MAHARASHTRA

As per the award of the tribunal, 0.25 MAF water has been allocated to Maharashtra. In order to avail this, the government of Maharashtra is required to harness water utilization schemes in the part of the Narmada basin falling under its territory. At present, these schemes are yet to be started. Maharashtra is getting 27% of the total power generated by the Sardar Sarovar Project. For this purpose, 400 kV DC transmission line has been laid from the project site in Gujarat upto the border of Maharashtra. Maharashtra has, on its part, constructed power-related structures on its border in order to avail of its share of power.

EMERGENCE OF OPPOSITION TO DAMS

An attempt to focus on the extensive measures to harness Narmada water led to unfavourable public reactions. It would be interesting to know how a very small protest against a big dam, nurtured by time and circumstances, flourished into a widespread, pervasive, and massive opposition.

During the year 1970, when survey and research studies had been conducted at the instance of the NWDT, cultivators in Nimar region of Madhya Pradesh started preliminary protests against Sardar Sarovar Dam proposed in Gujarat. They organized Nimar Bachao, Narmada Bachao (Save Nimar, Save Narmada) Samiti. The committee appeared before the tribunal on 28 October 1974 and expressed that it would not be appropriate to allow the fertile land of Nimar to get submerged. It also expressed its views on issues like displacement and other related matters. The committee argued that since 86.8% of the Narmada basin lies in Madhya Pradesh, and 99.5% water in Narmada comes from the catchment area lying in Madhya Pradesh, it is entitled to get 24.10 MAF water, while Gujarat should get only 0.75 MAF.

The committee was further of the view that since Gujarat needed power, it wanted to construct a high dam on Narmada and get the maximum amount of power generated thereby. This would submerge thousands of acres of fertile land of Nimar region. The representatives of the committee also submitted memorandum to then Prime

Minister Indira Gandhi and argued against the proposed height of the dam in Gujarat.

When a group of experts of the NWDT toured Nimar in February 1975, thousands of farmers, party workers, public organizations, representatives of gram panchayats, and local bodies raised a demand before the tribunal to save the fertile land of Nimar from submergence. When the NWDT passed its final award in 1979, widespread movements occurred in this part of Madhya Pradesh. Hundreds of farmers of their own volition participated in Nimar Bachao Andolan (Save Nimar Movement). About 5,000 people made a demonstration in Bhopal, the capital of Madhya Pradesh. Veteran Congress leaders Arjun Singh and Motilal Vora joined the demonstrators and got arrested. Virendra Kumar Sakhlecha, the then chief minister of Madhya Pradesh, announced the government's opposition to the NWDT award in the state legislative assembly. The chief minister also expressed his determination to get the award altered.

Hereafter, the political scenario of power in Madhya Pradesh underwent a change. With this change, the Nimar Bachao Andolan subsided for various reasons, including a crisis in the leadership of the movement and a lack of clarity of the objective. As a result of intense political activity and the dominance of Narmada issue at the national level, the government of Madhya Pradesh later accepted the award. Even though the Nimar Movement had subsided by then, the public awareness induced by agitation, both during the hearing conducted before the NWDT and after the award in organizing opposition thereto, turned the issue of Narmada into a topic of wide discussion and exchange of views. As a result, the Narmada issue moved out of the limited scope of 'Save Nimar' and became a live issue of concern and cogitation in wider circles.

In the year 1983, activists of a Delhi-based voluntary organization Kalpavriksh visited some areas of the Narmada Valley and published a critical report on the proposed major dams in the valley. A few social workers of Gujarat raised the issue of the rights of tribal communities being affected by the submergence of projects and the need for having major dams. In the year 1986, a voluntary organization called Setu (The Bridge) organized a seminar on the topic of 'existence'. Medha Patkar, an active worker of Setu, working in the tribal areas of Gujarat, was one of the organizers of this seminar. A member of the Delhi-based organization Marg (The Path), Vasudha Dhagamvar participated in this seminar and presented a paper dealing with the problems of displacement of people of the villages of Maharashtra likely to be affected by submergence. In this seminar, a detailed discussion was held on the after-effects of the proposed dam in Gujarat. Patkar, along with Vasudha Dhagamvar, visited the villages of Dhule district about which the latter had spoken in her presentation before the seminar.

During 1980–83, a few families of villages Sinduri and Manibeli of Maharashtra, which were first to be affected by submergence, were displaced. This made people at large concerned and aware about the issues of rehabilitation and resettlement. In the year 1985, the tribal villagers affected by submergence started organizing themselves. The main reason for this was the prevalent confusion and a host of misunderstandings with regard to dams, areas of submergence, and rehabilitation. The villagers of Dhule district of Maharashtra, who organized themselves due to fear of displacement, formed a committee on 16 February 1986 called the Narmada Dharangrast Samiti (Picketing Committee). The members of this committee started meeting officers to know about the issues of submergence, displacement, and rehabilitation. Initially, they attended district-level meetings called by the collector. During these meetings, they would ask officers to furnish detailed information about dams and the issue of submergence. The committee also visited the 16 sites, which were identified by the government of Maharashtra for developing as rehabilitation sites. The members of the committee expressed their dissatisfaction about the proposed rehabilitation sites.

The committee submitted a number of memoranda to Dhule district administration in the years 1986, 1987, and 1988. The main demand in these memoranda was to furnish complete information about dams, displacement, and rehabilitation. This committee got a survey done and collected detailed information about the affected people and their resources so that they could challenge the data of government survey and assessment. Right at this time, a contact happened to be established between the people of the tribal area of tehsil Alirajpur (now district) of Madhya Pradesh, who were going to be affected by the dam in Gujarat, and the Narmada Dharangrast Samiti of Maharashtra. Villagers of Alirajpur had also started their agitation since 1980, under the leadership of Khedut Mazdoor Chetna Sangath (KMCS), for their rights vis-à-vis issues about possible submergence.

In Madhya Pradesh, during 1986–88, a committee called the Narmada Ghati Nav Nirman Samiti was formed in the Nimar region with a view to highlight the issue of submergence. Important leaders devoted to the social service of Nimar like Kashiram Trivedi, Baijnath Mahoday, Prabhakar Mandlik, Phoolchand Patel, and Shobharam Jat played important roles in formulating this committee. In the year 1987, the committee, with a view to make people aware about the issue of submergence, organized Narmada Kalash Yatra (Journey with Narmada Water Pot) in the villages of Nimar and assembled on the banks of Narmada at Rajghat in Barwani.

In the middle of 1987, Medha Patkar contacted active workers of Narmada Ghati Nav Nirman Samiti at Barwani. Meetings between the members of this committee and workers from Madhya Pradesh, Maharashtra, and Gujarat continued to be organized at Rajghat (Barwani). These meetings served to prepare a framework of organization at each village and to induce awareness amongst people with regard to dam and related issues. The workers would visit village after village in Nimar region and organize people. Though the failure of Save Nimar Movement made the natives of Nimar indifferent to the cause, later the villagers got interested and enthusiastic about it. In the meantime, towards the end of the year 1987, the famous social worker Baba Amte organized a meeting on big dams at Anandwan, and expressed his deep concern about the ongoing activities in this connection.

In August 1988, the agitating groups of Maharashtra, Madhya Pradesh, and Gujarat came together and, by consensus, decided to oppose the dam and the resultant displacement. As a result of this decision, on 18 August 1988, rallies and demonstrations were organized in Dhule (Maharashtra), Barwani (Madhya Pradesh), and Kevadiya (Gujarat).

NARMADA BACHAO ANDOLAN (SAVE NARMADA MOVEMENT)

While concerns about submergence and displacement were being expressed in Gujarat, Maharashtra, and Madhya Pradesh, many support groups came into existence. These groups, in their own way, separately at their level, expressed their wrath against the proposed dam in Gujarat and displacement due to submergence. Starting from Dhule (Maharashtra) in the year 1987, many support groups have formed in Delhi, Pune, Indore, Bhopal, Ahmedabad, and Vadodara. Since 1990, support groups in favour of the Narmada agitation also began to be formed in West Bengal, Karnataka, Kerala, Tamil Nadu, Bihar, Rajasthan, and Saurashtra.

These support groups would exchange information, news, views, and analyses. Such exchanges induced these groups to carry out a collective campaign against government policies and projects pertaining to Narmada. After 1988, these different groups, sharing the same views and aims, collectively came to be identified as NBA. This name of collective identity owes its origin to the effect of Nimar Bachao, Narmada Bachao (Save Nimar, Save Narmada) Movement launched in the decade of the 1970s. Thus, all activities against the dam and displacement were subsumed in NBA.

The offer of World Bank to finance the Sardar Sarovar Project, and opposition thereto by the NBA, gave this movement international recognition and support. The activists of NBA established contacts at the international level in order to actively oppose the proactive role of the World Bank for construction of the dam. These contacts led to worldwide protests in America, Britain, Japan, Australia, Canada, Germany, France, and the Scandinavian countries by groups active in the field of environment in their respective countries against the World Bank.

Opposition to the dam acquired a new identity as the NBA brought a spurt in the agitation movements. On 22 February 1989, a rally against the dam was organized at the site of the Sardar Sarovar Dam. More than ten thousand people participated in the rally. On 4 April 1989, hundreds of people led by the NBA took part in the demonstration at Bhopal. On 28 September 1989, people in hundreds assembled in Harsud, which came under the submergence of Indira Sagar Project (Madhya Pradesh), and demonstrated against the submergence. On 6 March 1990, under the aegis of the NBA, thousands of people picketed for 28 hours on the Mumbai–Agra National Highway at Khalghat Bridge on Narmada. The same day, social worker Baba Amte left his residence in Anandwan (Maharashtra), came to Khalghat, and participated in the movement of the NBA. The villagers constructed a small cottage in village Chhoti Kasrawad near Barwani for his residence.

During 1990–91, picketing, demonstrations, and protest rallies under the aegis of the NBA continued to be organized in Maharashtra and Madhya Pradesh. The NBA opposed the offer of the World Bank to finance Sardar Sarovar Project and, as a result, the World Bank constituted an independent review committee known as Morse Committee. The Morse Committee submitted its report in June 1992. In the report, the committee recommended that World Bank should disassociate itself from the Sardar Sarovar Project. This recommendation was taken as a victory by the NBA. The NBA organized a victory rally in Kasrawad, Madhya Pradesh.

A meeting between the NBA and the government of Gujarat was organized for the first time on 12 September 1992 at Jamnagar for a direct exchange of views on the issue of the dam. During the discussion, people from Saurashtra led by the NBA put forward their questions related to the dam before the then chief minister of Gujarat and sought reply thereto. Later, such direct question-answer sessions were also organized at Rajkot, Bhavnagar, and Bhuj.

In the meantime, based on the recommendations of the Morse Committee, the World Bank prescribed certain conditional ties. Considering the obligation of fulfilling the conditional ties within a prescribed time frame, Government of India took a decision not to take any assistance from World Bank for the Sardar Sarovar Project and informed the bank accordingly. Thus, on 30 March 1993, the World Bank disassociated itself from the project. This was a great victory for the NBA.

In May 1993, the NBA launched satyagraha in Manibeli (Maharashtra), which would be the first village to be submerged by the Sardar Sarovar Dam. On 3 June 1993, Medha Patkar sat on an indefinite fast in Mumbai, demanding a complete review of the dam. After the declaration by the central government to carry out a comprehensive review of the dam, she concluded her fast on the fifteenth day. While an agitation for review of the dam was done on 11 July 1993, the house of one Bhulabhai Tarvi of village Vadgam, the first village of Gujarat to be affected by the dam, was submerged. Following this, the village of Manibeli was submerged. The people of Manibeli led by the NBA entered the water of Narmada chest-deep and refused to move out. The administration, in order to save their lives, had to use force to remove them.

In continuation of the protests in Manibeli, the NBA, on 17 July 1993, issued a warning of surrendering their lives to water till death. This was announced in the wake of the government's inability to take any action towards implementation of its declaration made in June 1993 to do a comprehensive review of the Sardar Sarovar Dam. After the threat, the government of India, on 5 August 1993, constituted a five-member review team under the chairmanship of Jayant Patil, member of the Planning Commission. On the constitution of the review team, the NBA withdrew its life threat.

During October–November 1993, the NBA undertook a journey called Narmada Yatra from Amarkantak to Sardar Sarovar Dam. The objective of the journey was to launch a campaign of public awareness against the construction of the dam.

On 16 March 1994 in New Delhi, Medha Patkar, along with 50 co-workers, raised slogans and thronged into the office of Kamal Nath, Minister of Environment and Forest. This ended with a discussion between the minister and the agitators, which lasted for two hours. Medha Patkar presented the cause of agitation against the dam. The union minister agreed to send a four-member team from the Ministry of Environment and Forests to Kevadiya to find out facts. On the very next day, that is, on 17 March 1994, the four-member team visited Sardar Sarovar Project site.

During the intervening night between 20 and 21 March 1994, the supporters of Sardar Sarovar Dam ransacked the office of the NBA at Dandia Bazar in Vadodara, Gujarat. The dam supporters threw away the furniture of the office and burnt some records as well. On 21

March, they gheraoed Medha Patkar for three hours and disrupted the road traffic. Gujarat Police safely extricated Patkar and her supporters and took them to an unknown place.

The NBA, under the leadership of Patkar, took out a massive rally against the dam on 18 February 1974 in Bhopal. Police arrested 741 demonstrators in New Delhi in this connection. The NBA resorted to picketing at Jantar Mantar against the dam from 19 April. On the seventh day of the picketing, then Prime Minister P.V. Narasimha Rao met a seven-member group of representatives of agitators and promised to get information about the facts. On his assurance, the picketing was called off.

On 20 May 1994, for the first time, the NBA filed a Public Interest Litigation (PIL) in the Supreme Court on the issue of rehabilitation and other related issues, and demanded to stay the construction of the Sardar Sarovar Project and the comprehensive review thereof. The Supreme Court bench comprising the former chief justices of India, Justice M.N. Venkatachaliah, Justice J.S. Verma, and Justice S.P. Bharucha, heard arguments of the government and the NBA and granted permission for listing the petition for hearing on 21 May 1994. The Supreme Court sent notices to Government of India and the governments of Madhya Pradesh, Gujarat, Maharashtra, and Rajasthan, and also to the NCA and Sardar Sarovar Narmada Nigam Ltd, asking for their feedback.

Jayant Patil Dam Review Committee, constituted by Government of India at the instance of the NBA, submitted its report to the government on 20 July 1994.

On 4 December 1994, the former chief minister of Gujarat Chhabildas Mehta, while participating in a programme sponsored by the organizations of edible oil producers and industrialists of Saurashtra, stated that he could prove that Medha Patkar, the leader of the NBA, took financial aid from foreign countries and created hurdles in the development of the country. Mehta further stated that if Patkar were to file a defamation suit against him, he would prove that in order to carry on disruptive activities in the country, Patkar had taken financial help from outside. Refuting this allegation, Medha Patkar asked the chief minister of Gujarat to tender an apology.

The issue of the NBA, taking foreign aid to organize agitations, was finally clinched in April 2007 after Government of India submitted an affidavit in a case filed on this subject in the Supreme Court. It was stated in the affidavit that in the enquiry on the question of taking foreign money, no case of violation of the provisions of Foreign Contribution (Regulation) Act (FCRA), either by the NBA or the institutions associated with it, was found. The central government had filed this affidavit in reference to a PIL filed on this subject.

On 21 November 1994, Medha Patkar along with her companions staged an indefinite hunger strike at Roshan Pura Square, Bhopal, opposing the Sardar Sarovar Dam. The hunger strike lasted for 26 days. Madhya Pradesh Vidhan Sabha was in session and the state government issued a statement, expressing concern for Medha Patkar. It declared that it would not support further construction of the dam. Following this, Digvijaya Singh, the then chief minister of Madhya Pradesh, went to Roshan Pura Square on 16 December 1994 and asked Medha Patkar to break her fast by offering a glass of milk.

In the meantime, the Supreme Court asked the central government to publish the Jayant Patil Review Committee report on the Sardar Sarovar Project. This 90-page report disclosed various observations. It was stated in the report that the addressing of environmental issues pertaining to the dam was indefinite, and that the number of persons estimated to be displaced was much less than those who would be actually displaced. There would be no sufficient land for all the displaced persons. It was also observed in the report that a law should be passed entitling the project-affected persons to have a stake in the profit accruing from the project.

In reference to the PIL filed by the NBA, the Supreme Court passed an interim order on 5 May 1995. It was stated in the order that the height of the dam would not exceed 80.3 metres till a decision was made in the case. In the meantime, news emerged from the chief engineer of the Sardar Sarovar Dam to raise the height of the dam from 80.3 metres to 85 metres. The NBA termed this as contempt of the interim order of the Supreme Court. In protest against this, the NBA organized a Nyay Yatra (Journey for Justice) on 11 October 1995 from Barwani (Madhya Pradesh) to New Delhi. Nyay Yatra reached Delhi on 19 October 1995. M. Reddy, Secretary to the Ministry of Water Resources, held discussion with the representatives of the NBA in Nyay Yatra. Followed by it, he assured the representatives that the height of the dam would be raised only after completing arrangements for rehabilitation and that the interim order of the Supreme Court would be respected fully.

Along with the protests against Sardar Sarovar Dam, the NBA also continued to organize regional rallies and protest marches between 1996 and 1997 in Madhya Pradesh against dams at Maheshwar, Bargi, Upper Beda, Lower Goi, and other places. On 11 January 1998, thousands of people led by the NBA assembled at the site of Maheshwar Dam and demanded a review of the project and even withheld the construction of the project. On 25 January 1998, six activists of the NBA resorted to hunger strike. The state government, under pressure, stopped work at the Maheshwar Dam and declared to constitute a task force for review of the dam. This task force submitted its report in October 1998. The unit, in its report, recommended to reanalyse the cost-benefit ratio studies along with some other recommendations. The government did not agree with the task force recommendations.

On 18 February 1999, during the hearing in the PIL filed by the NBA on Sardar Sarovar Dam, the Supreme Court withdrew the stay order on the construction of the dam passed in May 1995, and permitted to increase the dam height from 80.3 metres to 88 metres, including a three-metre high hump.

In order to press the government of Madhya Pradesh to implement the recommendations of the task force regarding Maheshwar Dam, the NBA undertook a 26-day fast in April 1999 at Bhopal. The state government agreed to stop work on the two other dams, but refused to stall work on the Maheshwar Dam. The outcome of this agitation was so profound that the two German companies, which had come forward to finance the construction of the Maheshwar Dam, withdrew themselves from the project.

THE HISTORIC JUDGMENT

The Supreme Court, after concluding the hearing in the PIL filed by the NBA, passed its final judgment on 18 October 2000. In the final judgment, the Supreme Court granted permission to construct the Sardar Sarovar Dam upto a full height of 138.68 metres (455 ft). The Supreme Court bench comprising Chief Justice A.S. Anand, Justice B.N. Kirpal, and Justice S.P. Bharucha, passed this judgment by a majority of 2:1. Justice Bharucha gave a dissenting judgment, calling for the stoppage of the construction of Sardar Sarovar Dam. This judgment gave a serious jolt to the NBA. Medha Patkar, the apex leader of the NBA, termed the Supreme Court judgment as anti-people and stated that their peaceful and non-violent movement would continue. The said judgment, which permitted the raising of the dam, imposed a condition of getting clearance from the Environment and Rehabilitation authorities before proceeding with the construction.

OPERATIVE PART OF THE SUPREME COURT JUDGMENT

Writ Petition (Civil) 319/94 (Judgment of 18 October 2000, Supreme Court of India)

While issuing directions in regard to the disposal of this case, two conditions needed to be kept in mind: (i) the completion of the project at the earliest, and (ii) ensuring compliance with the conditions on which the clearance of the project was given, including the completion of the relief and rehabilitation work and taking ameliorative and compensatory measures for environmental protection in compliance with the scheme framed by the government, thereby protecting the rights under Article 21 of the Constitution. Keeping these principles in view, the following directions were issued:

- Construction of the dam will continue as per the award of the tribunal.
- As the Relief and Rehabilitation Subgroup has cleared the construction up to EL 90 metres, the same can be undertaken immediately. Further raising of the height will be only pari passu with the implementation of the relief and rehabilitation and on the clearance by Relief and Rehabilitation Subgroup. This subgroup will give clearance of further construction after consulting the three Grievances Redressal Authorities (GRAs).
- The Environment Subgroup under the secretary of the Ministry of Environment, Forest, & Climate Change will consider and give, at each stage of the construction of the dam, environment clearance before further construction beyond EL 90 metres can be undertaken.
- The permission to raise the dam height beyond EL 90 metres will be given by the NCA, from time to time, after it obtains the above-mentioned clearances from the Relief and Rehabilitation Subgroup and the Environment Subgroup.
- The reports of the Grievances Redressal Authorities, and of Madhya Pradesh in particular, shows that there is a considerable slackness in the work of identification of land, acquisition of suitable land, and the consequent steps necessary to be taken to rehabilitate the project oustees. We direct the states of Madhya Pradesh, Maharashtra, and Gujarat to implement the Award and give relief and rehabilitation to the oustees in terms of the packages offered

by them, and these states shall comply with any direction in this regard which is given either by the NCA or the Review Committee or the Grievances Redressal Authorities.

- Even though there has been substantial compliance with the conditions imposed under the environment clearance, the NCA and the Environment Subgroup will continue to monitor and ensure that all steps are taken not only to protect but to restore and improve the environment.
- The NCA will within four weeks from today draw up an Action Plan in relation to further construction and the relief and rehabilitation work to be undertaken. Such an Action Plan will fix a time frame so as to ensure relief and rehabilitation pari passu with the increase in the height of the dam. Each state shall abide by the terms of the action plan so prepared by the NCA, and in the event of any dispute or difficulty arising, representation may be made to the Review Committee.
- However, each state shall be bound to comply with the directions of the NCA with regard to the acquisition of land for the purpose of relief and rehabilitation to the extent and within the period specified by the NCA.
- The Review Committee shall meet whenever required to do so in the event of there being any unresolved dispute on an issue which is before the NCA. In any event, the Review Committee shall meet at least once in three months, so as to oversee the progress of the construction of the dam and the implementation of the R&R programmes.
- If, for any reason, serious differences in the implementation of the Award arise, and the same cannot be resolved by the Review Committee, the Committee may refer the same to the Prime Minister, whose decision, in respect thereof, shall be final and binding on all concerned.

- The Grievances Redressal Authorities will be at liberty, in case the need arises, to issue appropriate directions to the respective states for due implementation of the R&R programmes, and in case of non-implementation of its directions, the GRAs will be at liberty to approach the Review Committee for appropriate orders.
- Every endeavour shall be made to see that the project is completed as expeditiously as possible.

This and connected petitions are disposed of in the aforesaid terms.

A.S. ANAND
Chief Justice of India
New Delhi

VERITABLE MILESTONES OF THE JUDGMENT

The Supreme Court, in its judgment, also removed the illusion that big dams are detrimental to the environment. With regard to big dams, paragraphs 124, 239, and 242 of the historic judgment of the Supreme Court are veritable milestones. They are reproduced here for ready reference.

Para 124:

'....The construction of a dam undoubtedly would result in the change of environment but it will not be correct to presume that the construction of a large dame like Sardar Sarovar will result in ecological disaster. India has an experience of over 40 years in the construction of dams. The experience does not show that the construction of a large dam is not cost effective or leads to ecological or environmental degradation. On the contrary, there has been ecological upgradation with the construction of large dams.'

Para 239:

'Since long the people of India have been deriving the benefits of the river valley projects. At the time of independence, foodgrain was being imported into India but with the passage of time and the construction of more dams, the position has been reversed. The large-scale river valley projects per se all over the country have made India more than self-sufficient in food. Famines which used to occur have now become a thing of the past. Considering the benefits which have been reaped by the people all over India with the construction of the dams, the Government cannot be faulted with deciding to construct the high dam on River Narmada with a view to provide water not only to the scarcity areas of Gujarat but also to the small areas of the State of Rajasthan where shortage of water has been there since time immemorial.'

Para 242:

'....Large dams can also become instruments in improving the environment, as has been the case of western Rajasthan, which transformed into a green area because of Indira Gandhi Canal which draws water from Bhakra Nangal Dam....'

Justice B.N. Kirpal
18 October 2000

After this milestone judgment, the opposition against big dams got totally diverted to the issue of displacement and rehabilitation arising therefrom. After 2000, the NBA continued its activities almost in every project area with regard to displacement and rehabilitation. Their tactics of picketing, agitation, and demonstration continued. The NBA filed a number of PIL petitions in Madhya Pradesh High Court on issues like displacement without rehabilitation, unsatisfactory compensations, and non-availability of basic facilities in rehabilitation sites. The NBA also filed a PIL petition with regard to rehabilitation concerning the Maan Project in the year 2004.

Similarly, with regard to the Indira Sagar and Omkareshwar projects, petitions were filed in the years 2005 and 2007 respectively. PIL petitions with regard to Upper Beda and Maheshwar projects on the issue of unsatisfactory rehabilitation were filed in the year 2008.

In the PIL petitions pertaining to the Indira Sagar and Omkareshwar projects, the Madhya Pradesh High Court in July 2009 passed an interim stay order on the construction of canals and the acquisition of land thereof. The high court also passed stay orders on the filling of the reservoirs of the two projects upto FRL. The government of Madhya Pradesh filed an appeal in the Supreme Court against the said stay orders of the high court. The Supreme Court, in its judgment dated 25 February 2010, permitted the resumption of the construction of canals in the two projects with a few conditionalities. In the judgment, a condition has been inserted that the work of command area development of the two projects would be done satisfactorily.

THE POLITICS OF WATER

Even though the NWDT has, in its award, finally decided that Gujarat would get 9 MAF water and that the height of the Sardar Sarovar Dam would be 455 ft (138.68 metres), the issue of the height of the dam continued to be a matter of cold war between Madhya Pradesh and Gujarat for long. During the proceedings before the NWDT, and after the final award, people's reaction was very combative in Madhya Pradesh with regard to the issue of submergence.

Save Nimar Agitation was launched for protecting the fertile land of Nimar by reducing the height of the dam. Considering the strong public sentiments, this agitation was supported by people's representatives at the regional and state level. The political parties and groups active in Gujarat and Madhya Pradesh to oppose the construction of big dams also supported this agitation.

In fact, during that time, the issue of submergence due to Sardar Sarovar Project had become a very effective instrument for garnering public support in Madhya Pradesh. No representative of the people worth one's name could get the support of the people, unless he/she opposed submergence. In these circumstances, it was difficult for the state to accept the award in the normal way, keeping the height of Sardar Sarovar at 455 ft.

The final award was passed by the NWDT in 1979. It drew sharp reactions in the Nimar belt of Madhya Pradesh. Hundreds of cultivators actively participated in Save Nimar Agitation. About 5,000 cultivators took part in the demonstrations in Bhopal. Senior leaders of Congress Party gave themselves up to be arrested. At that time, the Janta Party government was in power in Madhya Pradesh. Virendra Kumar Sakhlecha, the then chief minister of Madhya Pradesh, speaking in the state legislative assembly, declared the government's opposition to the award. When Arjun Singh became the new chief minister, the award was put before the state cabinet for approval. Shiv Bhanu Singh Solanki, a senior member of the cabinet, opposed the approval of the award plumped for sending the matter for review before the central government.

In 1987, the then chief secretary of Madhya Pradesh K.C.S. Acharya and Chief Engineer Matin Ahmad prepared a detailed note based on the technical facts and logic, submitted it to Arjun Singh, and proposed that the matter be taken up with Gujarat government for reducing the height of the Sardar Sarovar Dam. In the meantime, R.L. Gupta, a reputed irrigation engineer and secretary of the irrigation department, prepared a detailed treatise based on the facts and technical consideration in support of reducing the height of the dam. It was published in prominent newspapers of the region. The same year, the vice chairman of State Planning Board and the former State Minister of Irrigation, Ram Chandra Singhdeo, opposing the proposed height of the dam, publically declared that the award of the NWDT be reviewed.

The same year, Gujarat alleged that due to indifference of Madhya Pradesh, the clearance from Ministry of Environment and Forests for the construction of dam was not coming forth. In the middle of 1988, nearly 300 important personalities submitted a memorandum to the then prime minister Rajiv Gandhi, demanding a stay on construction of the dam and reconsideration of different aspects pertaining to Sardar Sarovar Dam.

On 6 March 1990, about 10,000 cultivators demonstrated on Bombay-Agra National Highway at Khalghat and stopped the traffic, whereupon the state government, which was formed only a day before under Chief Minister Sunder Lal Patwa, declared that the government of Madhya Pradesh would ask the central government for a review of Sardar Sarovar Dam. On 5 April 1990, the chief minister of Maharashtra, Sharad Pawar, announced in the state legislative assembly that no village in Maharashtra would be submerged as per the accepted plan. On 18 May 1990, the then prime minister V.P. Singh assured a delegation, who had demanded the reduction of the dam height, that the central government would get all aspects of Sardar Sarovar Dam reviewed.

On 6 January 1994, at New Delhi, the chief minister of Madhya Pradesh Digvijaya Singh stated in a press conference that he would discuss the matter with the chief ministers of the concerned states for reducing the height of Sardar Sarovar Dam. He further stated that a note has been prepared in this regard after discussion with three senior engineers. This statement created turmoil in Gujarat. The government of Gujarat requested the prime minister for his intervention in the matter. The view of Gujarat was that there was no justification, whatsoever, for the question of raising the height after having resolved all issues and work on the project having been started.

Sensing the controversy, Prime Minister Narasimha Rao called a special meeting and discussed the matter with all concerned. Commenting on the statement of Digvijaya Singh, the interim chief minister of Gujarat Chhabildas Mehta stated that Gujarat would not agree to reduce the height of the dam even by an inch, and that the government of Gujarat would not hold any further discussion in this regard.

An all-party meeting was called in Madhya Pradesh on 21 February 1994 for getting the height of the dam reduced. A consensus emerged in this meeting that pleaded to reduce the height of the dam to 436 ft in place of 455 ft. Chief Minister Digvijaya Singh was authorized to persuade the concerned states to agree to this. In the meantime, on 23 February 1994, the then Central Minister for Water Resources, Vidya Charan Shukla, met the members of the parliament from Gujarat at New Delhi. He assured them that the height of Sardar Sarovar Dam would not be brought down. Sushil Chandra Varma, member of parliament and former chief secretary of Madhya Pradesh, stated that the award of the NWDT is final with regard to the Sardar Sarovar Dam. The state government was obfuscating the people by raising the question of height.

In Madhya Pradesh, there was no let-up in the activities pertaining to the reduction in height. On 3 March 1994, while replying to the debate on the address of the governor in the legislative assembly of Madhya Pradesh, Chief Minister Digvijaya Singh made a fervent appeal to the people of Gujarat to reconsider the height of the Sardar Sarovar Dam so that the fertile land and 30,000 families in Madhya Pradesh could be saved. On 6 March 1994, the chief minister met Prime Minister Rao and requested him to intervene in the matter. The chief minister shared the observation of the prime minister that it would be congenial to all if the concerned chief ministers agree to it.

In the meantime, the Central Minister for Environment and Forests Kamal Nath supported the demand of Madhya Pradesh for reducing the height. He stated that the only impact of the reduction of height would be on the production of power, and that Madhya Pradesh would be willing to cede its share of power. In answer to a question of a correspondent, Kamal Nath said that the NWDT award was not a Gita or a Bible which could not be amended.

In the Gujarat legislative assembly, the interim chief minister Chhabildas Mehta stated that Gujarat would not accept any amendment with regard to Sardar Sarovar Dam. All the four states are bound by the award, and no single state can take any decision in the modification thereof. The chief minister and the minister for Narmada affairs, Babu

Bhai Patel, stated that the award could not be reviewed before 45 years. The general secretary of All India Congress Committee, Ahmad Patel, on the other hand, sent a telegram to the prime minister to ignore the demand of Madhya Pradesh regarding the height of the dam. The ex chief minister of Gujarat, Amar Singh Chowdhary, stated in Gandhi Nagar that the height of the dam would not be reduced and that the indulgence of statements being issued in this connection would be seriously dealt with. Prabodh Rawal, president of Gujarat Pradesh Congress Committee, also criticized the assertion of Madhya Pradesh to reduce the height of the dam.

On 9 March 1994, the Central Minister for Water Resources, Vidya Charan Shukla, gave an assurance in the Lok Sabha that there was no change in the opinion of the central government with regard to the award. Amid controversy arising out of statements in Madhya Pradesh for the reduction in the height of the dam, and anti-statements by Gujarat, Prime Minister Rao, after a meeting of the Congress party in New Delhi, advised Digvijaya Singh to desist from making any new statement that might aggravate the situation.

Senior leaders of Gujarat Congress met the prime minister and requested that the committee appointed by the Ministry of Environment for Sardar Sarovar Project be dissolved, and that the demand of Madhya Pradesh for reducing the height of the dam be ignored. Leaders of BJP from Gujarat also made similar demands before the central government. The opposition leader in Gujarat legislative assembly, Keshubhai Patel, in a press conference in Vadodara on 23 March 1994, demanded the dissolution of the committee appointed by the Ministry of Environment and alleged that the central government, under the pressure from those opposed to the dam, was playing with the sentiments of the people of Gujarat.

In May 1994, a demonstration in opposition to Sardar Sarovar Dam was organized in London before the office of the Indian High

Commission. Katherine Goldsmith, wife of Edward Goldsmith, environmentalist and writer in England, took part in the demonstration.

Madhya Pradesh, despite facing opposition from Gujarat and the adverse attitude of the central government, continued with efforts for reducing the height of the dam. Chief Minister Singh discussed the matter with members of the parliament from Madhya Pradesh on 2 August 1994 at New Delhi. He informed them that by reducing the height of the dam, 25,000 acres of fertile land could be saved from submergence and 32,000 families could be saved from displacement. Commenting on these activities of Madhya Pradesh, the chief minister of Gujarat Chhabildas Mehta made a statement on 4 September 1994 at Gandhi Nagar that Gujarat would not agree to any proposal for reducing the height of the dam.

On 3 December 1994, Subhash Yadav, the then deputy chief minister and minister of the NVDA in Madhya Pradesh, announced the constitution of two committees of the members of legislative assembly. Announcing this in the state legislative assembly, he said that these committees would study rehabilitation works and displacement caused by the Sardar Sarovar Project. The two committees, one headed by MLA Vikram Varma and the other headed by Ramchandra Singhdeo, studied the areas coming under submergence by Sardar Sarovar Project and the condition of the displaced persons of Madhya Pradesh resettled in Gujarat. The committee, headed by Singhdeo, stated in its report that Gujarat had not done justice to the displaced persons. Neither were they rehabilitated in a better way, nor were any means of livelihood provided to them.

In a meeting of the SSCAC convened on 4 January 1995 at New Delhi, Naresh Narad, vice chairman of the NVDA in Madhya Pradesh, demanded that the construction of the dam should be stopped till those getting displaced till June 1995 were properly rehabilitated. It was demanded that the height of the dam be reduced from 455 ft to 436 ft. The governments of Gujarat and Maharashtra rejected both of these demands by Madhya Pradesh. Looking at the objections by Madhya Pradesh, the NCA asked the rehabilitation subgroup to submit its report. Sticking to its demand, on 17 January 1995, Madhya Pradesh filed an affidavit in the Supreme Court for reducing the height of the dam to 436 ft. It was stated in the affidavit that even the NWDT had concluded that for the purpose of irrigation, it was enough to keep the height of the dam at 436 ft.

The chief minister of Gujarat Keshubhai Patel addressing a press conference at New Delhi on 24 March 1995, alleged that Medha Patkar, stationed in Madhya Pradesh, was organizing agitation against the interests of Gujarat. He said that the attitude of the government of Madhya Pradesh was negative regarding the construction of the dam. The chief minister further stated that there was no question of reducing the height of the dam. Environment and Forests Minister Kamal Nath stated on 11 April 1995 at Bhopal that non-compliance of conditionalities imposed at the time of accord for Sardar Sarovar Dam would lead to discontinuation of the construction of the dam. The conditionalities could be complied only if the height of the dam was reduced. In another instance, the chief minister of Maharashtra Manohar Joshi met a delegation led by Medha Patkar during his visit to Indore. He assured the delegation that the government of Maharashtra would not file any document in the Supreme Court without consulting them in the petition, thus opposing the construction of the dam filed by the NBA. The chief minister of Maharashtra gave assurance of full cooperation to the NBA.

Chief Minister Patel and the leader of the opposition in the legislative assembly Amar Singh Chowdhary met Prime Minister Rao on 24 April 1995. They complained that because of the obstructionist attitude of Madhya Pradesh, the entire work of the project was going behind schedule. Both the leaders also submitted a copy of the resolution unanimously passed by the Gujarat legislative assembly stating that the people of Gujarat would not accept, under any circumstance, the height of Sardar Sarovar Dam to be less than 455 ft. In continuation of the efforts by Madhya Pradesh for reducing the height of the dam, on 30 May 1995, members of the parliament from Madhya Pradesh led by the chief minister presented a memorandum in this connection to Prime Minister Rao. The very next month, in a press conference organized at New Delhi on the occasion of completion of 111 days in the office of his government, Patel stated mouthfully that he would not allow the interests of Gujarat to be sacrificed at the instance of the displaced and the environmentalists of Madhya Pradesh. The chief minister further stated that the sanctioned height of the dam would not reduce under any circumstance.

In November 1995, the government of Madhya Pradesh constituted a committee of five members of the parliament, headed by Dileep Singh Bhuria, to inspect rehabilitation works done by Gujarat. The committee comprised of Suraj Bhan Solanki, Radhakishan Malviya, Ajit Jogi, and Rameshwar Patidar. After visiting the concerned areas, the committee in its report expressed dissatisfaction over the rehabilitation work done by Gujarat. In the report, it was recommended that the government should examine the possibility of determining the minimum height of the dam.

On this issue, a meeting of representatives of Rajasthan, Gujarat, Maharashtra, Madhya Pradesh, and the central government, under the chairmanship of the Central Minister of Water Resources Janeshwar Mishra was convened on 5 July 1996 at New Delhi. In the meeting, the chief minister of Madhya Pradesh alleged that Gujarat was not fulfilling its obligations with regard to rehabilitation. Madhya Pradesh, on this basis, demanded the reduction of the dam height from 455 ft to 436 ft. But the other three states emphasized on keeping the height as per the direction given in the award.

The meeting of the review committee of the NCA, called on 15 July 1996, to preside over the dispute of the dam height ended without reaching any decision. Later, Janeshwar Mishra and the representatives of the four states met Prime Minister H.D. Deve Gowda and discussed the matter with him. On 16 July 1996, the prime minister discussed the issue of the dam height with the chief ministers of Madhya Pradesh, Rajasthan, Maharashtra, and Gujarat, namely Digvijaya Singh, Bhairon Singh Shekhawat, Manohar Joshi, and Suresh Mehta respectively. It was agreed that for the time being, the height of the dam be kept at 436 ft and, during the ensuing five years, the availability of water in the dam should be studied after which the decision about raising the height to 455 ft shall be taken. Initially, it was treated as a victory for Madhya Pradesh. When this news was published, Dr Sharad Jain, ex-chairman of NVDA in Madhya Pradesh, stated at Bhopal that a distorted version of the decision taken at the prime minister's meeting was being presented. The fact was that no decision was taken for reducing the height of the dam to less than 455 ft.

In the meantime, the NCA approved the proposal to raise the height of the Sardar Sarovar Dam to 110 metres. The chief minister of Gujarat Shankersinh Vaghela, during his press conference on 15 November 1996 at Ahmedabad, expressed his happiness about the decision and stated that despite the objections by Madhya Pradesh, Gujarat would start work on raising the height of the dam with effect from 15 December 1996.

In continuation of these events, the chief minister of Madhya Pradesh declared on 3 December 1996 at New Delhi that in order to get the height of the dam reduced, he would soon visit Ahmedabad, Vadodara, and Rajkot. Reacting on this, the former chief minister of Gujarat Keshubhai Patel stated that if the chief minister of Madhya Pradesh attempts to visit Gujarat with the object of getting the dam height reduced, he would not be allowed to enter. Retorting to this statement, Digvijaya Singh stated that nobody could restrain him from visiting Gujarat. Such gambits were frequently resorted to by Gujarat and Madhya Pradesh. Such attacks and counter-attacks, however, tapered off by the day. After the Supreme Court's decision in 2000, stating that the Sardar Sarovar Dam would be raised to its full height of 455 ft (138.68 metres), the politics of water between Gujarat and Madhya Pradesh came to an end.

IRRIGATION AND POWER BENEFITS TO PEOPLE

The major projects built or are under construction on Narmada and its tributaries have started yielding benefits of irrigation and power to the people of the valley. The projects now under construction would substantially add to the present irrigation potential on their completion.

The major projects which have been completed, under construction, or to be taken up in future would provide irrigation to a very large area of land in Gujarat, Madhya Pradesh, and Rajasthan. The statewise details are as follows:

IRRIGATION FROM MAJOR PROJECTS

From the following major projects, a total of 35,18,235 hectares of agricultural area would be under irrigation in Gujarat, Madhya Pradesh, and Rajasthan.

Irrigation from Major Projects

Project	Irrigation Potential Created (In Hectares)	Beneficiary Districts
Gujarat		
Sardar Sarovar	18,62,000	Bharuch, Narmada, Vadodara, Dahod, Godhra, Anand, Nadiad, Ahmedabad, Gandhinagar, Mehsana, Patan, Banaskantha, Bhavnagar, Surendra Nagar, Rajkot, Kutch

Project	Irrigation Potential Created (In Hectares)	Beneficiary Districts
Rajasthan		
Sardar Sarovar	2,46,000	Jalore, Barmer
Madhya Pradesh		
Indira Sagar	1,23,000	Barwani, Khargone, Khandwa
Omkareshwar	1,47,000	Khargone, Dhar, Khandwa
Rani Avanti Bai Sagar	1,57,000	Jabalpur, Narsingpur
Bargi Diversion	2,45,000	Jabalpur, Katni, Rewa, Satna
Upper Narmada	18,616	Dindori, Anuppur
Punasa Lift	35,000	Khandwa
Chinki	70,800	Narsingpur, Raisen
Maan	15,000	Dhar
Shahid Chandra Shekhar Ajad (Jobat)	9,848	Dhar
Barna	60,750	Raisen, Sehore
Sukta	16,599	Khandwa
Tawa	2,42,915	Hoshangabad, Harda
Kolar	33,593	Sehore
Matiyari	10,120	Mandla
Upper Beda	9,917	Khargone
Lower Goi	13,760	Barwani
Halon	13,040	Mandla
Upper Budner	9,800	Mandla
Atariya	1,986 and 9,000 by feeder canal from Bargi Diversion	Jabalpur, Katni
Sher, Shakkar, Machhrewa	64,700	Narsingpur, Chhindwara
Dudhi	50,586	Hoshangabad
Morand/Ganjal	52,205	Hoshangabad, Harda
Total	**35,18,235**	

IRRIGATION FROM MEDIUM PROJECTS

Of the 135 medium projects mentioned in the master plan, 20 projects have been completed as of January 2014. The total irrigation potential created is 65,062 hectares. The irrigation and beneficiary districts are as under:

Irrigation from Medium Projects

S. No.	Project	Irrigation Potential Created (In Hectares)	Beneficiary Districts
1	Banjar	2,567	Mandla
2	Bichhiya Tank	2,797	Mandla
3	Dhuandhar Tank	1,410	Mandla
4	Mohgaon Tola Tank	2,094	Jabalpur
5	Barnu Tank	2,030	Jabalpur
6	Madai Tank	2,564	Jabalpur
7	Bahoriband Tank	5,423	Katni
8	Choral Nakheri	3,900	Indore
9	Dukrikheda	7,625	Hoshangabad
10	Sampan	7,486	Betul
11	Paras	1,507	Dewas
12	Jalgaon	546	Balaghat
13	Dejla Dewda	9,000	Khargone
14	Kunda	1,938	Dhar
15	Satak	2,707	Khargone
16	Segwal	1,803	Barwani
17	Sakalda Tank	2,060	Manawar
18	Chandra Keshar Tank	7,605	Dewas
19	Pariyat Tank	Drinking water 20.30 MCM	Jabalpur
Total		**65,062**	

IRRIGATION FROM MINOR PROJECTS

Of the 3,000 minor projects included in the master plan of Madhya Pradesh, 1,032 projects have been completed up till January 2014. These projects have created a potential of irrigating 1,94,165 hectares of land. The cultivators are also irrigating additional 2,67,000 hectares by lifting water from these minor projects by means of pumps.

DEVIATION FROM MASTER PLAN

Madhya Pradesh was allocated 18.25 MAF of water after the claim put by the state before NWDT. The claim was strongly supported by a master plan. This master plan includes 29 major, 135 medium, and about 3,000 minor irrigation projects. After the declaration of the award, the state finally scanned the proposed projects before taking them up for construction.

Thus, in many proposed projects, there came a need to deviate partially or fully from the original project. By and large, the construction of major irrigation projects is being followed as per the master plan. In case of medium and minor projects, it is not followed because of the fact that in the preceding years, after the declaration of the award, new, feasible, and technically suitable proposals of medium and minor projects came to the fore. The state government focused on utilizing the allocated water with the best proposals of medium and minor projects available at hand.

IRRIGATION IN THE PERIPHERAL LAND OF RESERVOIRS

Under large dam reservoirs, irrigation can be provided in the area on its right and left banks through gravity flow. Large areas adjoining the periphery of the reservoir remain unirrigated because the gravity flow system does not work in these areas. In Madhya Pradesh, an innovative scheme has been prepared to extend irrigation on the boundaries of reservoirs. Under the scheme, electric lines would be erected on the periphery of the reservoirs. The cultivators in the periphery would take electric connection from these lines and, by means of pumps, may lift water for irrigation. The scheme is called Garland Project because electric poles and power lines along the periphery seem like a garland. The Garland Project has been taken up on the periphery of Indira Sagar reservoir in Khandwa. A total of 350 transformers have been installed on the periphery of the reservoir and electric lines have been extended. The cultivators of 128 villages on the periphery of the reservoir in Khandwa would be able to irrigate about 19,000 hectares of agricultural land by lift irrigation. If this project succeeds, it will be extended to other reservoirs of Narmada Valley in Madhya Pradesh.

CREATION OF IRRIGATION POTENTIAL

It is imperative to know that the mere construction of the irrigation project and creation of infrastructure for targeted irrigation potential does not fulfil the sole object of any project. The objective of the construction of the project is actually fulfilled only when the built-up irrigation potential is entirely or mostly used for agriculture. It has been seen in practice that despite an irrigation potential of a major project having been created, a very small ratio of that potential is used for a few years.

Thus, difference subsists between the created irrigation potential and the actual irrigation done. There are reasons for this difference, one of the main reasons being that the cultivators who are traditionally used to rain-fed agriculture are not adapted in doing irrigation through canals. Nor are they eager for immediate crop rotation in keeping with the availability of canal water. For taking water to the field, creating minor canals, water channels, and water courses depends on the nature of the soil. Even after construction of canal network, if the land to be irrigated is not levelled, and if appropriate field channels are not made therein, the irrigation potential of the project cannot be utilized fully.

It also takes time to procure improved variety of seeds, appropriate chemical fertilizers and insecticides, and improve agricultural implements in keeping with the requirement of canal irrigation. Hence, the cultivators would not be able to fully utilize the available water. This is why it takes about five to seven years for the irrigation potential of the major and the medium projects to be utilized to the maximum extent.

With the cost of construction of major projects increasing day by day, there is now a perceived realization to take advantage thereof at the earliest. Now, even in the last phase of construction of the project, the development work of the command area and training of cultivators is taken up. Madhya Pradesh and Gujarat have taken steps to ensure partnership of cultivators in the management of irrigation. Under this programme, water utilization committees are elected at the village, block, and district levels. Members of these committees are given training in the maintenance of canals and water management. Training programmes for cultivators are also organized. In Madhya Pradesh, a large number of *kolaba sahayak*s (outlet assistants) have been appointed for controlling and ensuring judicious distribution of water in the command area of the distribution network of major projects. The *kolaba sahayak* controls the water flow as per the availability of water and requirement of the cultivator.

The participatory irrigation management under the official act Madhya Pradesh Sinchai Prabandhan Me Krishkon Ki Bhagidari Adhiniyam, 1999 has been enforced in Madhya Pradesh. This has brought positive results. In two major irrigation projects (Maan and Jobat), the full irrigation potential is being utilized by the cultivators. These two projects set an example of utilizing the full potential (25,000 hectares), that too by the second year of completion of the project, with active participation of cultivators in water management activities.

HYDROELECTRICITY

The three major hydroelectricity projects based in Narmada basin have, since their inception till the years 2013–14, generated 65,180.62 million units of electricity. The details of the project-wise annual generation of power are charted as follows:

Annual generation of power from 3 major hydroelectric projects (in million units)

Year	Sardar Sarovar	Indira Sagar	Omkareshwar
2003–04	—	196.51	—
2004–05	255.288	1331.85	—
2005–06	1897.949	2573.36	—
2006–07	3539.511	2605.58	—
2007–08	4358.030	2729.83	702.04
2008–09	2268.373	1568.67	799.78
2009–10	2447.780	2118.47	952.89
2010–11	3512.011	2197.76	998.88
2011–12	4270.25	3286.81	1376.87
2012–13	3609.22	2896.47	1264.27
2013–14	5701.30	4078.74	1633.13
2014–15	2842.63	2562.00	1132.00
2015–16	2095.10	1974.28	958.61
2016–17	3132.41	3320.79	1427.71
2017–18	1111.46	881.68	443.57

Madhya Pradesh, Maharashtra, and Gujarat have received power from Sardar Sarovar Project (till the years 2013–14) as per their share:

Madhya Pradesh : 57%, that is, 18,119.16 million units
Maharashtra : 27%, that is, 8,602.12 million units
Gujarat : 16%, that is, 5,097.55 million units

MICRO, MINI, AND SMALL HYDROPOWER POTENTIAL

The hydroelectricity projects have been classified on the basis of their capacity for generation of power, shown as follows:

Classification of hydroelectricity projects on the basis of their capacity for generation of power

Micro Hydroelectricity Project (Micro Hydro)	upto 0.1 megawatt (100 kW)
Mini Hydroelectricity Project (Mini Hydro)	from 0.1 to 2 megawatt (101–2,000 kW)
Small Hydroelectricity Project (Small Hydro)	from 2 to 25 megawatt (2001–25,000 kW)
Major Hydroelectricity Project (Major Hydro)	above 25 megawatt (above 25,000 kW)

Megawatt is a unit of energy-generation capacity. If any hydroelectricity project with a capacity of one megawatt is run for one hour with full load capacity, then 1,000 units of power are generated.

Madhya Pradesh has constituted a new department in October 2010 to exploit micro, mini, and small hydel potential of all the river basins in the state, including the Narmada basin. The department, known as New and Renewable Energy Department, shall take up the hydropower projects in all the river basins upto a potential of 25 megawatt.

FINANCIAL RESOURCES FOR THE PROJECTS

It is the primary responsibility of the states to make financial arrangements for the projects in the Narmada Valley where they are being constructed. Besides providing funds from their own financial sources for the construction of projects, the states also get financial assistance from Government of India under related programmes and credit from the loaning agencies. In the following section, we discuss the financial sources for the Narmada Valley projects, which are:

- States' own financial sources
- Accelerated Irrigation Benefit Programme (AIBP) of the government of India
- National Bank for Agriculture and Rural Development (NABARD)

STATES' OWN FINANCIAL SOURCES

The state governments, besides providing money from their state budgets, also exploit other possibilities. The government of Gujarat has incorporated a separate company by the name Sardar Sarovar Narmada Nigam Ltd. This company collects money through share capital. For Indira Sagar and Omkareshwar projects, the government of Madhya Pradesh created a joint venture with NHPC named NHDC. The share of government of Madhya Pradesh and NHPC in this venture is 49% and 51% respectively.

RESOURCES OF GOVERNMENT OF INDIA

Government of India provides financial assistance to such irrigation projects that are proposed or under construction in the state and are able to provide irrigation in a period of four years. The amount released to states under this programme (AIBP) are in the form of grants. For irrigation project in the general areas, the assistance under the programme is given to the extent of 25% of the estimated cost for the projects. In the scheduled areas and notified drought-prone areas, it extends financial assistance to the extent of 90% of the project cost. The financial assistance under this programme has been availed for the construction of Sardar Sarovar Project canals in Gujarat, Narmada canal in Rajasthan, and almost all projects in Narmada Valley in Madhya Pradesh.

NATIONAL BANK FOR AGRICULTURE AND RURAL DEVELOPMENT (NABARD)

The NABARD provides loans for the construction of irrigation projects. Generally, the states avail the grant under the AIBP, and the balance amount is taken as loan from the NABARD. During 2009-10, the rate of interest on loan from NABARD was 6.5%. This rate of interest is changeable as per the directions of the Reserve Bank of India. The rate of interest applicable for the years 2012-13 is 7.25%. The NABARD does not pay the amount of loan sought by the states in advance, but the states are required to spend the amount on projects in advance.

The NABARD reimburses the amount already spent as loan. Based on the position of and the benefit of irrigation to accrue from the project, the NABARD sanctions loan to the extent of 90% to 95% of the project cost. The states are required to repay the debt taken from the NABARD in predetermined installments.

INTERSTATE FINANCIAL LIABILITY

In the NWDT award, the cost of construction of Narmada Valley Development Project (NVDP) has been directed to be shared by the states having a share in Narmada. The interstate financial liability has been laid down as follows:

- The cost of construction of Indira Sagar Dam in Madhya Pradesh shall be shared to the extent of 17.63% by Gujarat.
- The cost of construction of the dam and powerhouse, that is to say, the energy production component of Sardar Sarovar Project in Gujarat shall be shared by Madhya Pradesh, Maharashtra, and Gujarat in the ratio of 57%, 27%, and 16% respectively. It may be mentioned that the electricity generated by this project shall be shared in the same ratio by the three states.
- The cost of construction of the dam and other related construction, that is to say, the irrigation component of the Sardar Sarovar Project shall be shared by Gujarat and Rajasthan in the ratio of 18:1.
- The cost of construction of 485.318 km long main canal from Sardar Sarovar Dam shall be shared by Gujarat and Rajasthan. The sharing of the cost shall be done on the basis of the flow of water (cubic ft.) per second in respect of 9 MAF and 0.5 MAF water allocated to Gujarat and Rajasthan respectively. In other words, the sharing liability of Rajasthan has been fixed on the flow of 0.5 MAF water in Narmada canal from Sardar Sarovar Dam to Rajasthan. In addition, for any change in gradient in the design of Narmada canal, approved by the NCA, the entire expenditure incurred shall be borne by Rajasthan.
- For calculating the financial liability in regard to Sardar Sarovar Project between Madhya Pradesh, Maharashtra, and Gujarat with

respect to the construction of energy production component, and between Gujarat and Rajasthan with regard to the construction of the irrigation component, the expenditure is deemed as stated. Out of the total expenditure incurred on the dam and related works (unit-1) of Sardar Sarovar Dam, 43.9% shall be deemed as expenditure on the irrigation component and 56.1% thereof shall be deemed on electricity component.

- The government of Gujarat shall, besides bearing expenditure on displacement and rehabilitation of the people of Gujarat affected by the submergence, also bear the entire cost of displacement and rehabilitation of the affected people in Madhya Pradesh and Maharashtra. However, with regard to this liability, Gujarat has come out with certain arguments as per its interpretation of the award.

COST OF SARDAR SAROVAR PROJECT AND SHARES OF STATES

The actual cost of construction of Sardar Sarovar Dam and the related constructions as per the rates in 2007–08 works out to be Rs. 3,504.81 crore. As per the award of the NWDT, Gujarat has contributed Rs. 545.68 crore to Madhya Pradesh as its 17.63% share in the cost of Indira Sagar Project. Gujarat has added this amount to the cost of construction of Sardar Sarovar Dam and its other related works. Thus, the total cost of the dam and related works (unit-1) adds up to Rs. 4,050.49 crore.

The total cost of construction of the powerhouse (including civil work and electrical equipment), that is to say, the energy production component of the project, works out to be Rs. 2,952.07 crore.

For calculating the share of Madhya Pradesh, Maharashtra, and Gujarat in the energy production component of the project, the expenditure on energy production has been assumed to be 56.1%

of Rs. 4,050.49 crore, which happens to be the total cost of the dam and related works. This works out to be Rs. 2,272.32 crore. Adding this amount to Rs. 2,952.07 crore, which is the cost of construction of the powerhouse of the project, including the cost of electrical equipment, the total cost of energy production component of the project works out to be Rs. 5,224.39 crore. According to the award, the shares of Madhya Pradesh, Maharashtra, and Gujarat in this cost work out as stated:

Shares of states in the cost of Sardar Sarovar Project

Madhya Pradesh	57%	Rs. 2977.90 crore
Maharashtra	27%	Rs. 1410.59 crore
Gujarat	16%	Rs. 835.90 crore
Total		Rs. **5224.39 crore**

According to the award, Gujarat has borne expenditure on displacement and rehabilitation of the people not only of its state but also those of Madhya Pradesh and Maharashtra, whose land and people are affected by Sardar Sarovar Project. Under this head, the total expenditure on rehabilitation works out to be Rs. 4,887.59 crore.

As per the award, Gujarat and Rajasthan have to bear expenditure of the total cost of irrigation component of the project in the ratio of 18:1. For the purpose of calculating this cost, the total expenditure on the irrigation component is assumed to be 43.9% of Rs. 4,050.59 crore, which is the total cost of the dam. The amount worked out to be Rs. 1,778.16 crore. Rajasthan had accordingly contributed Rs. 93.59 crore, as per its laid-down ratio for this cost.

COST OF NARMADA CANAL IN GUJARAT

A cumulative expenditure of Rs. 5,864.84 crore has been incurred on Narmada main canal works, and an expenditure of Rs. 13,267.00 crore has been incurred on branch, sub-branch canals, and distributaries in Gujarat till 31 March 2013.

COST OF NARMADA CANAL IN RAJASTHAN

A cumulative expenditure of Rs. 1,966.55 crore, including Rs. 670.75 crore, towards the payment of shared cost to Gujarat was incurred.

COST OF PROJECTS IN MADHYA PRADESH

Initially, the construction of major projects in Narmada Valley was done under the Water Resources Department of the state. Later, a new department of Narmada Valley Development (NVD) and NVDA was created. The construction of major projects in the Narmada Valley was brought within the purview of the authority. The projects completed before the creation of the NVDA are Tawa, Barna, Kolar, Sukta, and Matiyari. The NVDA has so far constructed Maan, Shahid Chandra Shekhar Azad (Jobat), and Punasa Lift Irrigation projects. The construction of Indira Sagar and Omkareshwar project dams and powerhouses has been completed through NHDC. The canals of these projects are being constructed by the NVDA. Besides these, Upper Beda and Lower Goi projects, Bargi Diversion Canal, and Rani Avanti Bai Lodhi Sagar left-bank canal are being constructed by the NVDA.

For the construction of Bargi left-bank canal, the government of Madhya Pradesh has taken loan from NABARD. Of the total estimated cost of Bargi Diversion Project, 75% is met by taking loan from NABARD, and the remaining 25% is obtained from Government of India under the AIBP as grant. For the construction of canals of Indira Sagar and Omkareshwar projects, assistance by way of grant under the AIBP has been received to the extent of 90%. In the same way, AIBP financial assistance is received to the extent of 90% for Upper Beda, Lower Goi, and Punasa Lift projects.

The request to declare Bargi Diversion Project as a national project is pending with the central government. Details of the estimated cost of the projects under construction and those completed in Madhya Pradesh are stated as follows. The cost escalation is likely to be for the projects under construction and those proposed by the time they are completed.

Cost of projects in Madhya Pradesh

S. No.	Name of the project	Total cost of the project (in crores)
Completely Constructed Projects		
1	Tawa	113.00
2	Barna	18.90
3	Kolar	120.00
4	Sukta	12.60
5	Matiyari	30.00
6	Maan	246.00
7	Shahid Chandrasekhar Azad (Jobat)	230.00
8	Rani Avanti Bai Sagar Dam and Powerhouse	429.82
9	Rani Avanti Bai Canal Head Powerhouse	52.00
10	Indira Sagar Dam and Powerhouse	4355.57
11	Indira Sagar Canal Head Powerhouse	61.61
12	Omkareshwar Dam and Powerhouse	2224.73
Under Construction (as of Dec 2018)		
13	Rani Avanti Bai Sagar (Left-Bank Canal)	1515.00
14	Bargi Diversion (Right-Bank Canal)	5127.00
15	Upper Beda	224.00
16	Lower Goi	360.00
17	Punasa Lift	488.00
18	Indira Sagar Project canals	3182.77
19	Omkareshwar Project canals	2504.80
20	Halon	414.00
Proposed Projects		
21	Upper Narmada	684.00
22	Upper Budhner	300.00
23	Ataria	130.00
24	Chinki	578.00
25	Sher, Shakkar, and Machhrewa	647.00
26	Dudhi	506.00
27	Morand and Ganjal	522.00

SUBMERGENCE BY LARGE WATERBODIES

When a parental river is supported by constructing a dam, the water so stored badly disturbs human settlement in the upstream area of the dam. A large dam requires a large area to be a reservoir, whereas the submergence due to medium and minor dams is limited. Besides human settlement, submergence also affects agriculture and forest areas to a large extent in major dam projects. For this reason, major dam projects in Narmada Valley face opposition not only from the people affected, but also from environmentalists and social organizations. Uprooting thousands of families from their place and resettling them away from the reservoir area has been a very difficult task for the government. But it has been more difficult and painful for the families who have to unwillingly move away from their age-old home and environment. With respect to land coming under submergence, it is worth noting that the clearance for the construction of projects is given only if the command area of the project is about ten times the area coming under submergence.

In case of projects where hydroelectricity is also to be generated along with irrigation, permission is granted only when the proposed command area is ordinarily ten times the area coming under submergence due to storage of water for irrigation. With regard to forestland coming under submergence, twice this area should be brought under compensatory afforestation. Of the projects so far constructed in Narmada Valley, the largest submergence has been caused by Sardar Sarovar, Indira Sagar, Omkareshwar, and Rani Avanti Bai Lodhi Sagar projects. The effects of submergence due to prominent major projects in the Narmada Valley have been listed as follows:

Submerged land (in hectares) in Sardar Sarovar Project

Type of submerged land	Madhya Pradesh	Gujarat	Maharashtra	Total
Private land	8,246	1,877	1,519	11,642
Forest land	2,731	4,165.91	6,488.54	13,385.45
Government land	10,208	1,069	1,592	12,869
Total	21,185	7,112	9,599	37,896.45

Submerged villages in Sardar Sarovar Project

State	No. of affected villages			No. of affected families (approx.)
	Wholly	Partially	Total	
Madhya Pradesh	1	177	178	23,614
Gujarat	3	16	19	4,740
Maharashtra		33	33	4,195
Total	4	226	230	32,549

AFFECTED DISTRICTS

The district of Vadodara in Gujarat is affected by submergence. Within Vadodara, 11 villages of Nadod taluka, 4 villages of Nasvadi taluka, and 4 villages of Chhota Udaipur taluka are affected by the reservoir. In Madhya Pradesh, the districts affected by the reservoir are Alirajpur, Dhar, Barwani, and Khargone. Within the districts, 26 villages in Alirajpur, 77 villages in Dhar, 65 villages in Barwani, and

10 villages in Khargone are affected by submergence. In Maharashtra, 33 villages in Dhule have been affected.

Submerged land/villages/families in Rani Avanti Bai Project

Type of land	Submerged land (hectare)	No. of affected villages			No. of affected families
		Wholly	Partially	Total	
Agricultural land	14,872				
Forest land	8,478	22	140	162	5,474
Government land	3,447				
Total	16,797				

The districts affected by submergence are Jabalpur (19 villages), Mandla (95 villages), and Seoni (46 villages).

Submerged land/villages/families in Indira Sagar Project

Type of land	Submerged land (hectare)	No. of affected villages			No. of affected families
		Wholly	Partially	Total	
Private land	44,741	156+1 town (Harsud)	95	252	44,631
Forest land	40,332				
Other land	6,653				
Habitat (*abadi*) land	779				

The districts affected by submergence are Khandwa (165 villages), Harda (48 villages), and Dewas (39 villages).

Submerged land/villages/families in Omkareshwar Project

Type of land	Submerged land (hectare)	No. of affected villages			No. of affected families
		Wholly	Partially	Total	
Private land	3,520.087	5	26	31	6,290
Forest land	5,277.000				
Other land	1,273.000				
Habitat (*abadi*) land	60.642				
Total	10,130.729				

The districts affected by submergence are Khandwa (22 villages) and Dewas (9 villages).

Other Projects

The following table lists the details of submerged land in respect of other completed, ongoing, and proposed major projects:

Submerged land in respect of other completed, ongoing, and proposed major projects

S. No.	Project	Submerged land (in hectare)	Status of project (December 2018)
1	Tawa	20,055.44	Completed
2	Barna	7,700	Completed
3	Kolar	2,503	Completed
4	Sukta	1,206	Completed
5	Matiyari	678.46	Completed
6	Maan	1,111.30	Completed
7	Shahid Chandra Shekhar Azad (Jobat)	1,301.94	Completed
8	Upper Beda	1,144.65	Ongoing
9	Maheshwar	4,856	Completed
10	Upper Narmada	2,952	Proposed
11	Upper Budhner	6,153.84	Proposed

S. No.	Project	Submerged land (in hectare)	Status of project (December 2018)
12	Halon	1,838.35	Ongoing
13	Atariya	1,028	Proposed
14	Chinki	5,937.84	Proposed
15	Sher	2,256.20	Proposed
16	Machhrewa	505.87	Proposed
17	Shakkar	2,121	Proposed
18	Dudhi	3,849	Proposed
19	Morand	1,940	Proposed
20	Ganjal	109.20	Proposed
21	Lower Goi	1,328	Ongoing
22	Rosra	3,066.34	Proposed
23	Basania	2,116.10	Proposed
24	Raghavpur	2,553	Proposed

THE TRAUMA OF DISPLACEMENT

No trauma could be more painful for a family than to get uprooted from the place where it has lived for generations. The displaced are not only uprooted, but asked to resettle at a place where it may be harder for them to adjust. To build a major irrigation dam project, the displacement of a large population from the area of the future reservoir is an unavoidable act. The uprooted families face social, cultural, and economic problems along with lots of hardship and distress in life, only because they are surrendering their homeland for a cause which holds the promise of progress and prosperity. The displaced families, in fact, make great sacrifice on their part for the country and the society at large.

The issues of displacement, resettlement, and rehabilitation of the uprooted people have been very sensitive and challenging. More than one lakh families have been subjected to displacement from their ancestral places and forced to be resettled at various places out of the reservoirs areas. The states of Gujarat, Madhya Pradesh, and Maharashtra have developed various resettlement sites in their area for the displaced families of the respective states.

Madhya Pradesh had to bear the maximum brunt of this challenge, as most of the dams in the valley were constructed here. The maximum area under submergence by Sardar Sarovar Dam in Gujarat also falls in Madhya Pradesh. In this situation, besides the construction of dams, the work of resettlement and rehabilitation of a large chunk of the affected population was done in the state.

The Sardar Sarovar Dam built in Gujarat is 214 km long with a maximum width of 16.1 kms. The FRL is 138.68 metres (455 ft) and the Maximum Water Level (MWL) is 140.21 metres (460 ft). The water retention capacity of the Sardar Sarovar Dam is 4.73 MAF. The shape, size, and water retention capacity of the reservoir have entailed the submergence of a large chunk of land in Madhya Pradesh, Gujarat, and Maharashtra. A total of 30,734 families were living in the reservoir area, which amounts to approximate 3.50 lakh people. The maximum affected families numbering 21,808 were in Madhya Pradesh. In Gujarat, 4,740 families, and in Maharashtra, 4,195 families were affected by the dam reservoir.

The NWDT has given detailed directions regarding the acquisition of land and property in Madhya Pradesh, Gujarat, and Maharashtra, which is affected by Sardar Sarovar Dam. These directions cover the acquisition of affected properties, provision to give residential plots, and basic civic amenities for resettlement and rehabilitation of the affected families in the rehabilitation sites. There is a provision in these directions that if any family of Madhya Pradesh and Maharashtra affected by the Sardar Sarovar reservoir intends to settle in Gujarat, the government of Gujarat would allow it and extend all benefits, as per its rehabilitation policy, to such families. The families who intended to settle down only in Madhya Pradesh and Maharashtra, as the case may be, would be resettled in the respective states as per the provisions of the rehabilitation policies of the respective states.

The total number of villages affected by submergence due to the Sardar Sarovar Dam in the three states is 244. Of these, only three villages in Gujarat, that is, Gadher, Badgaon, and Mukhari, and one village Rohana in Madhya Pradesh would be wholly and permanently submerged. Rohana in Kukshi tehsil in Dhar, Madhya Pradcsh, is uninhabited.

For proper rehabilitation of the families affected by the submergence caused by the dam, the states of Madhya Pradesh, Gujarat, and Maharashtra have formulated their own rehabilitation policies. It may be mentioned that the rehabilitation policies of the three states are more liberal than the arrangements made in the decisions and the final orders of the NWDT. While carrying out rehabilitation work in its entirety, the three states, as per requirement, have further liberalized the provisions from time to time.

The NWDT, along with directions for the construction of the dam on Narmada at Navagam (Sardar Sarovar), has laid down exhaustive guidelines for resettlement and rehabilitation of *abadi* (inhabited people) affected by the submergence. These guidelines provide that the acquisition of land for Sardar Sarovar Dam shall be done under the provisions of the Land Acquisition Act 1894. The government of Gujarat shall reimburse the expenditure to Madhya Pradesh and Maharashtra, which was incurred by them for land acquisition and resettlement of the displaced families. For the land acquired, Gujarat will pay land revenue to the states of Madhya Pradesh and Maharashtra as per the provisions of their respective land revenue codes. Similarly, temples and monuments of historic, religious, and archeological importance affected by the submergence of the reservoir would be shifted to new sites, and the expenditure involved shall be borne by the government of Gujarat.

The NWDT passed its award in the year 1979. In the light of the award, the actual work of rehabilitation started after a considerable delay. The resettlement and rehabilitation work pertaining to the Sardar Sarovar Project actually started in the year 1992. By this time, the cost of living in all related fields had increased manyfold. In this situation, it became necessary to bring about incremental revision in all related heads pertaining to rehabilitation works. Thus, guided by the directions of the award, Gujarat, Madhya

Pradesh, and Maharashtra have formulated their own shifting and rehabilitation policies.

RESETTLEMENT IN MADHYA PRADESH

The main provisions of resettlement and rehabilitation provide compensation for submerged agricultural land and houses, land for submerged agricultural land, and special rehabilitation package as options, besides financial assistance to displaced landless families, adult sons, and adult unmarried daughters of displaced families. Residential plots in developed resettlement sites are also provided along with various grants and facilities.

The families affected by Sardar Sarovar Project numbered nearly 21,808. Out of these families, 5,535 have opted to settle in Gujarat. The remaining 16,273 families have been resettled in Madhya Pradesh. In order to resettle these families, 88 rehabilitation sites in the command area of Narmada Valley have been developed.

In order to extend additional irrigation facility to the resettled families, the Kathora Lift Irrigation has been constructed in Kasravad tehsil, Khargone district. This lift irrigation system is designed to irrigate 6,945 hectares land. The benefits of the irrigation have started to show up since 2011.

RESETTLEMENT IN GUJARAT

In order to appropriately carry out the resettlement and rehabilitation of families affected by the Sardar Sarovar Project, the government of Gujarat constituted Sardar Sarovar Resettlement Agency in the year 1992. This agency has acquired its responsibility in resettling the affected families according to the rehabilitation policy of the state.

Resettlement of the affected families in Gujarat

Category	Families	Rehabilitation site
Affected families of Gujarat resettled in Gujarat	4,744	110
Affected families of Madhya Pradesh resettled in Gujarat	5,535	108
Affected families of Maharashtra resettled in Gujarat	772	18

Thus, 236 rehabilitation sites have been developed in Gujarat for resettling a total of 11,051 families affected by Sardar Sarovar Project.

RESETTLEMENT IN MAHARASHTRA

3,094 families in Maharashtra have been affected by Sardar Sarovar Dam. To resettle these families, nine rehabilitation sites have been developed in the state.

REHABILITATION AND DISCONTENTMENT

Madhya Pradesh submitted Action Taken Report (ATR) to the NCA about the completion of displacement and rehabilitation of the families likely to be affected on raising the height of the Sardar Sarovar Dam from 110.64 metres to 121.92 metres. Based on the ATR, the NCA took the decision to elevate the height of the dam upto 121.92 metres on 8 March 2006. From the very next day, Gujarat started work on increasing the dam height.

The NBA opposed this move. It was of the view that the claim by Madhya Pradesh on displacement and resettlement was false, and that the stipulations about rehabilitation of the displaced had not been complied with. About 350 activists of the NBA started picketing on 17 March 2006 at New Delhi, outside the Ministry of Water Resources. On the ninth day of the agitation, that is, on 25 March 2006, a delegation led by Medha Patkar met Prime Minister Manmohan Singh and lodged a complaint about the rehabilitation work in Madhya Pradesh pertaining to Sardar Sarovar Project. In the memorandum submitted to the prime minister, the demands that were raised included: the various aspects of the dam be reviewed by the Planning Commission, the work of raising the dam height be stopped, and the claim about rehabilitation work been carried out be examined.

As time went by, with no initiative being taken by the prime minister in this connection, Medha Patkar, along with the NBA workers, launched an indefinite hunger strike on 29 March 2006 at Jantar Mantar in New Delhi. Amid hunger strike and picketing, Saifuddin Soz, Minister of Water Resources, announced on 4 April 2006 that, at the instance of the prime minister, a group of ministers would visit Madhya Pradesh, Gujarat, and Maharashtra to inspect the rehabilitation work. Besides Soz, Meira Kumar, then Minister of

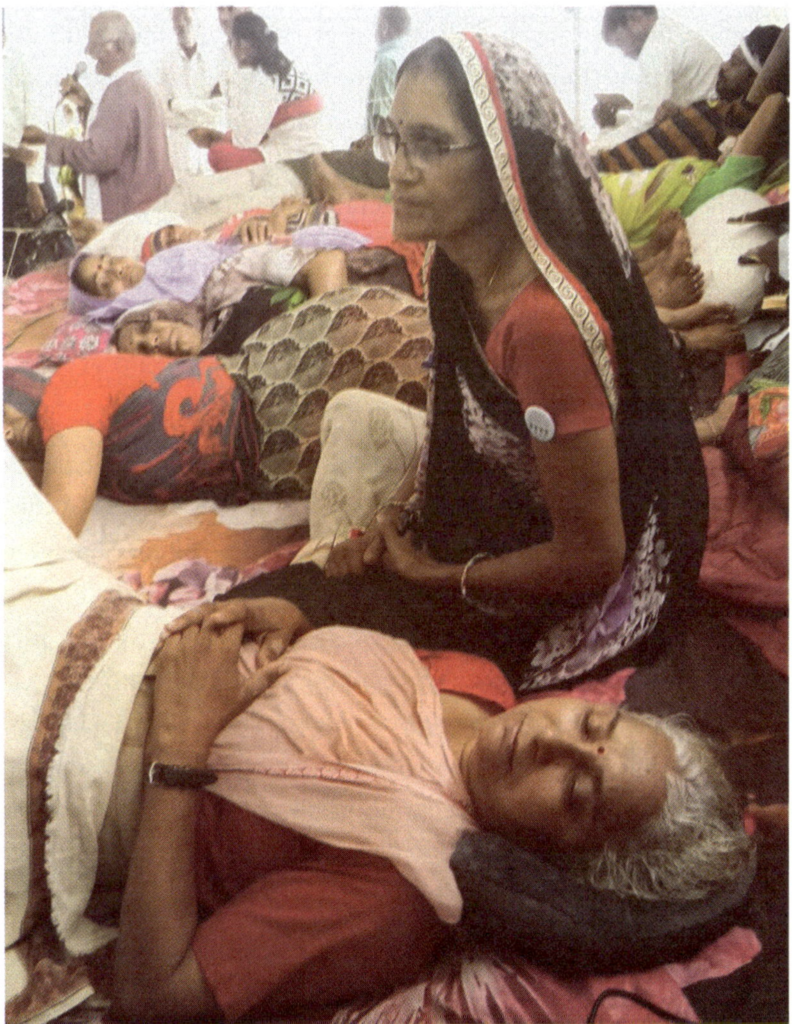

NBA leader Medha Patkar on hunger strike at Jantar Mantar, New Delhi, in April 2006

Social Justice and Empowerment, and Prithviraj Chavan, Minister of State in Prime Minister's Office, were included in the group. After this announcement, Soz offered lime water to Medha Patkar and her companions on the hunger strike and made them break the fast.

However, it was stated on behalf of the NBA that the fasting by Medha Patkar and others would continue until the team of ministers complete the tour and make their report public. On the eighth day of the indefinite hunger strike, that is, on 5 April 2006, late in the night, Delhi police arrested Medha Patkar and got her admitted in the emergency ward of All India Institute of Medical Science.

The group of ministers, on the other hand, came to Indore on 6 April 2006 from New Delhi to undertake a tour of the area. On 7 April, the group visited a few rehabilitation sites in the districts of Dhar and Barwani affected by the Sardar Sarovar Project. During the tour of the rehabilitation areas, these ministers were gheraoed by local people. At Dharampuri, the enraged villagers took out the key from the vehicle of Saifuddin Soz and raised slogans. After coming back to Indore, the group of ministers, while interacting with the media, expressed dissatisfaction over the displacement and rehabilitation work pertaining to Sardar Sarovar Project done in Madhya Pradesh and levelled many allegations. The group did not discuss the matter with the chief minister or any officer of the state. After halting for the night at Indore, the group of ministers left for New Delhi the next day.

In New Delhi, the activists of the NBA continued their picketing at Jantar Mantar. On the twelfth day of the hunger strike, Medha Patkar wrote a letter from the hospital to the prime minister, demanding that the ministers who toured the Narmada Valley make their report and views public. Meanwhile, the group of ministers met Prime Minister Singh and submitted their report.

In Madhya Pradesh, Chief Minister Shivraj Singh Chouhan called a press meeting on 19 April 2006 at his residence and took strong exception to the visit of the group of ministers and their hastily prepared report. The chief minister stated that despite there being an elected government in Madhya Pradesh, it was fully unconstitutional on the part of the group of ministers to tour areas affected by submergence without taking advice of the state government. The chief minister further stated that the tour created confusion among the people, and the ongoing work of rehabilitation had been badly affected. The chief minister in this connection wrote a letter to the prime minister and gave point-by-point explanation on the allegations raised in the report. Madhya Pradesh thus totally rejected all allegations pertaining to the rehabilitation work levelled by the group of ministers.

The Minister of Water Resources called a meeting of the review committee of the NCA on 15 April 2006 at New Delhi. All the chief ministers of the four states participated in the meeting. Since no decision could be reached in the meeting, the matter was forwarded to the prime minister for taking the decision.

Realizing the seriousness of the matter, Prime Minister Singh, on a visit to Germany on 24 April 2006, constituted a committee headed by the ex-comptroller and auditor general V.K. Shunglu to enquire into the rehabilitation work done in Madhya Pradesh. This was informed to the press in Berlin by the prime minister's media advisor. The other two members of the three-membered Shunglu Committee were the convener of the NGO Loksatta, Dr J.P. Narayan, and former vice chancellor of Jawaharlal Nehru University, Dr G.K. Chadha.

The chairman V.K. Shunglu and member Dr Chadha came to Bhopal on 19 May 2006 to discuss the matter with the chief secretary Rakesh Sahni, vice chairman of the NVDA Udai Verma, and other senior officers. For rendering necessary help to the committee, an advance team of National Sample Survey Organization (NSSO) reached Indore. The rest 100 members of the team reached Indore on 21 May. After training, the members of the survey team at Indore were sent to Dhar and Barwani districts. V.K. Shunglu informed the chief secretary that 100% survey of the displaced families who have been rehabilitated would be carried out. In the first phase, Dhar and Barwani, and in the second phase, Khargone and Jhabua districts would be surveyed. Chief Minister Chouhan also held discussion with the Shunglu Committee.

During the survey, the Shunglu Committee did not take any help from the NVDA or the state government; it completed house-to-house survey of 25,000 families in 177 villages and assessed the actual status of rehabilitation.

On 3 July 2006, V.K. Shunglu submitted his report to the Central Minister of Water Resources, Saifuddin Soz. The same day, Soz submitted the report to the prime minister. The Shunglu Committee found the ATR of the government of Madhya Pradesh pertaining to the displaced families factual and, by and large, accepted it. The prime minister took the decision of continuing the work of Sardar Sarovar Dam on the basis of the report, and submitted the papers of this decision before the Supreme Court.

MONITORING REHABILITATION WORK

The NWDT, in its final order, had directed to create mechanisms for executing and monitoring its decisions. Accordingly, the NCA had been constituted for the purpose by the Ministry of Irrigation. Under the NCA, various subgroups have been formed to execute and monitor decisions of the award.

The Rehabilitation and Resettlement Subgroup (R&R subgroup) of the NCA supervises the implementation of decisions/directions taken by the NWDT pertaining to rehabilitation in the three states. This subgroup is headed by the secretary of Ministry of Social Justice and Empowerment. The subgroup comprises the representatives of the three states, rehabilitation and resettlement experts of academic institutions, independently working social activists, and anthropologists and reputed NGOs working in the field of rehabilitation.

Through its rehabilitation units, the R&R subgroup of the NCA supervises and inspects rehabilitation work, develops rehabilitation sites in the three states, and transports affected families to the rehabilitation sites. In 2006, the NCA further strengthened the rehabilitation units by constituting six inspection teams. These teams regularly monitor and inspect the rehabilitation work.

Officers of the NCA also visit states and regularly inspect and supervise rehabilitation work. Mainly, these officers assess the status of implementation of decisions by the R&R subgroup, allotment of residential plots on rehabilitation sites, extension and standard of basic amenities in the rehabilitation sites, and issue necessary instructions to the concerned states.

INDEPENDENT ASSESSMENT BY THE STATES

Besides the monitoring of rehabilitation work by the NCA R&R subgroup, the three states also engage independent agencies for conducting study and supervision of rehabilitation work. The states of Madhya Pradesh, Gujarat, and Maharashtra have got the study of rehabilitation work of Sardar Sarovar Project done by academic institutions working in the field of sociology and anthropology. These institutions, from time to time, conduct a survey of villages affected by the submergence from socio-economic and cultural points of view, and appraise the states concerned with their decisions. On the basis of these reports, the states come to know in time about the problems, lapses, and difficulties with regard to the rehabilitation work. This information helps in taking a practical view while deciding about matters pertaining to resettlement and rehabilitation.

The main purpose of these surveys and studies is to assess the pre-displacement and post-rehabilitation status of the affected families from the socio-economic and cultural perspectives. The survey is mainly focused on the following 13 indices of human development:

- Specialties of the habitat
- Individuals' right to hold property
- Resources of joint property
- Economic composition
- Indebtedness
- Health
- Availability and quality of water sources
- Literacy and school activities
- Available services and provisions within the reach of the people

- Social organizations and contacts/relations
- Relationship with administrations
- Relationship of the rehabilitated group with the previous populace
- Caste-wise mapping of the area

REDRESSAL OF GRIEVANCES

Many issues were raised in the PIL filed by the NBA in the Supreme Court in the year 1994 against the construction of Sardar Sarovar Project. The issues of redressal of grievances of the families affected by the submergence were also included in them. The Supreme Court passed a verdict on this petition of the NBA on 18 October 2000. This was the first judgment of the Supreme Court on the directions and arrangements mentioned in the final award of the NWDT. This judgment contains directions for making separate arrangements in each state concerned for the redressal of grievances of the affected families. In this context, as desired by the Supreme Court, the states of Gujarat, Madhya Pradesh, and Maharashtra have each constituted GRAs in their states.

Gujarat had constituted GRA on 17 February 1999, even before the said judgment of the Supreme Court. The first chairman of the GRA in Gujarat was retired chief justice P.D. Desai, who had been the chief justice of Bombay, Calcutta, and Himachal Pradesh high courts. Gujarat has created a grievance redressal cell in the Sardar Sarovar Rehabilitation Authority, which is supervised by and works under the GRA. This cell works under the system of a single window arrangement and ensures that the applications of the displaced persons are heard and disposed of within the given time limit.

Madhya Pradesh also constituted a GRA on 30 March 2000 under the chairmanship of the ex chief justice of Patna High Court, Justice G.G. Sohani. Maharashtra constituted a GRA on 17 April 2000 under the chairmanship of Justice S.P. Kurdukar, a retired judge of the Supreme Court.

THE SAGA OF DISPLACEMENT OF HARSUD

Except Sardar Sarovar, all other major dam projects in Narmada Valley were taking shape one after another in Madhya Pradesh. Rani Avanti Bai Sagar Dam on Narmada was completed in 1988. During the completion period, the families displaced by the project were given compensation and grants under provisions of the Land Acquisition Act and the Land Revenue Code. For subsequent projects, a rehabilitation policy was formulated in the year 1985 by the NVDA. This policy was amended in 1989 and was made more liberal and extensive. In the case of Indira Sagar, Omkareshwar, Maan, Jobat, and other projects of the state, the benefits of rehabilitation were extended to the displaced under the provisions of this policy.

The rehabilitation provision applicable to Narmada Valley reservoirs in Madhya Pradesh is almost similar to the rehabilitation provisions applicable in case of Sardar Sarovar Project. The only difference is that for Sardar Sarovar Project, in case more than 25% of the land belonging to a landholder is submerged, there is provision for additional rehabilitation package. This additional package is given for purchasing a minimum of two hectares of agricultural land, whereas in the case of other projects, special rehabilitation grant is given on land actually acquired. Another comparison to Sardar Sarovar Project is that a displaced family of the project who declines to take a plot is given Rs. 50,000 in lieu thereof, whereas oustees of other projects in a similar situation is paid Rs. 20,000 in lieu of the plot.

The work of displacement and resettlement of affected families pertaining to major dams in Narmada Valley was a big challenge to Madhya Pradesh. The most challenging among these was the shifting and resettling of the township of Harsud from the area of submergence of Indira Sagar Dam. Harsud was a well-established town since British days, which, after independence, became a centre of trade and agriculture for the adjoining rural hinterland. Attempts for shifting the township of Harsud continued for a long time. As the height of Indira Sagar Dam started going up, the pressure for shifting of Harsud increased on the state government. It was generally believed that the government would have to bear the brunt of anger and strident public opposition on the shifting of Harsud. The public representatives of the area were also in quandary because their support base, being subjected to the unnatural act of displacement, went counter to their political interests.

In the year 2003, when relocating the town of Harsud became inevitable, the then chief minister Digvijaya Singh had a discussion with the dam constructing agency NHDC. The engineers of NHDC suggested that since Harsud was located on the last fringe of Indira Sagar Dam, it can be saved by constructing a protective guide wall between the reservoir and the township.

Agreeing with this proposal, on 7 January 2003, the state cabinet approved the proposal to construct the protective guide wall. In compliance with this proposal, the officers of the NHDC were directed to do the survey. Accordingly, a team of officers in NHDC went to Harsud. They perhaps thought that this proposal would be welcome in Harsud, but contrarily, the survey team had to face stiff opposition from the citizens of Harsud. This strident opposition to the survey indicated that the citizens of Harsud would want to shift and resettle at a new place. A further enquiry of this issue revealed that before the proposal of the protective wall came up, the survey of immoveable properties and calculation of compensation had already been prepared. Moreover, notices under relevant sections of the Land Acquisition Act had also been served to the owner of the property. Thus, every family was aware about the amount of compensation one would get for one's property.

The real issue for the opposition was to procure additional compensation apart from what was available under the rehabilitation policy in force. Since there was no provision for any special compensation, a situation of pressure and denial between people of Harsud and the NVDA continued for some time. In the meantime, the then vice chairman of the NVDA, Pradeep Bhargava, formulated a special rehabilitation package for Harsud and submitted it for approval before the chief minister.

On 18 January 2004, the then chief minister of Madhya Pradesh, Uma Bharti, went to the dam site Narmada Nagar to inaugurate the first unit of hydroelectricity of Indira Sagar Project. On this occasion, a delegation from Harsud met the chief minister and discussed the demands pertaining to the rehabilitation and relocation of the town of Harsud. Ultimately, on 14 May 2004, the chief minister granted approval to the proposal of the NVDA for a special rehabilitation package in Harsud. On knowing about the special rehabilitation package from newspapers dated 15 May, a wave of happiness swept over the entire township of Harsud. It was provided in the package that the families shifting from Harsud before 30 June 2004 would get 10% of the award amount, subject to a minimum of Rs. 25,000 and a maximum of Rs. 5,00,000, as a special city compensatory allowance. Immediately thereafter, the provision was amended, and those getting compensation upto Rs. 5,00,000 were sanctioned city compensatory allowances to the extent of 20%, subject to a minimum of Rs. 35,000 (those getting more than Rs. 5,00,000 as compensation would get 10% as city compensatory allowance, subject to a maximum of

Rs. 5,00,000). Besides, it was further provided that the families holding less than 540 sq. ft as residential plot in Harsud would get a plot of 540 sq. ft, families holding a plot between 540 sq. ft and 1,000 sq. ft would get a residential plot of 1,500 sq. ft, and the families holding a plot of more than 1,000 sq. ft would get a plot of 2,400 sq. ft at the new rehabilitation site.

The citizens of Harsud welcomed this package and made up their mind to shift. Since the last date for availing the package was declared to be 30 June 2004, there was a spurt in shifting. By this time, the work for providing necessary facilities in the new rehabilitation site at Chhanera was still in the process of being fully developed. The construction of these facilities was going at a slow pace. Problems arose when people started shifting from Harsud in a big way and started demanding basic amenities in Chhanera.

Chief Minister Uma Bharti entrusted the responsibility of overseeing the development work at Chhanera to the Minister of Public Works Department, Kailash Vijayvargiya, the Minister for Water Resources Department, Anoop Mishra, and the Chief Secretary, B.K. Saha. Both the cabinet minister and the chief secretary immediately called for a meeting of the senior officers of various departments at Chhanera on 27 June 2004 and took several decisions for accelerating the pace of the development work. This meeting was attended by almost all the principal secretaries, commissioners, and directors of development departments. It was clearly stated that for development works in Chhanera, the principal secretary of the concerned department would be directly responsible. Action on the various decisions taken in this meeting started the very next day. The senior officers remained present on the spot and got the construction and development work of the concerned department done. The officers and officials of the service department went from door to door of the resettled families and provided them services like ration card, healthcare, electric connection, etc.

In almost no time, about 95% of families demolished their houses, salvaged the useful material, and, along with household goods, shifted to the new rehabilitation site. Free transportation service having been sanctioned, the services of hundreds of trucks and tractors were made available. The effect of attractive rehabilitation package was such that many families, without waiting for the government vehicles, transported their goods on their own. In this manner, the town of about 35,000 population got resettled at village Chhanera at a distance of 12 kms from Harsud on Harsud–Khandwa road. This rehabilitation site at Chhanera is known as New Harsud. Ward numbers 9, 12, and 13 of Old Harsud were not affected by the submergence, and the property therein was not acquired. It seemed strange that leaving these three wards, the entire population shifted to a new location. Later, the government took the decision of acquiring these wards as well. Thus, by August-September 2004, the entire township of Harsud, having shifted from its original location, became a township of ruins and broken buildings.

One year after the displacement of Harsud, on 30 May 2005, the citizens observed the first 'death anniversary' of Harsud. A ceremony was organized at Gandhi Square of the dilapidated township, where hundreds of people from New Harsud participated. These people visited the ruins of their houses and refreshed their memories associated with their houses and town. It was a moving sight. The ceremony was attended by senior journalists from Delhi, like Rahul Dev, Ram Bahadur Rai, Ramsharan Joshi, and Gyan Ranjan, director of Jagran Institute of Management and Mass Communication (JIMMC), Ajay Upadhyay, politician and poet/composer Vithalbhai Patel, and writer Nirmala Bhuradia. On the occasion, a book entitled *Harsud, 30 June*, written by journalist Vijay Manohar Tiwari, was buried in a grave-like trench near Vijaya Stambha of Gandhi Square.

The rehabilitation site of Old Harsud at Chhanera was almost inhabited in a planned way. In subsequent years, the families who, after leaving Harsud, had temporarily gone to the cities of Khandwa, Indore, and other nearby towns, came back and settled in New Harsud. Presently, New Harsud, equipped with modern urban amenities and facilities, has developed into a complete, modern township. Old Harsud was located between Khirkiya and Talwadiya railway stations. The railway track, 52 kms in length, which lay between Khirkiya and Talwadiya, was affected by the submergence due to Indira Sagar Project. Hence, it was diverted and connected to New Harsud. While Old Harsud's flourishing trade and enterprise depended on dozens of nearby villages, in New Harsud, the lack of similar arrangements for trade and enterprise has led to the serious problem of unemployment.

GRIEVANCE REDRESSAL MECHANISM

On the lines of GRA for Sardar Sarovar Project, the government of Madhya Pradesh had, on 30 August 2001, constituted a separate GRA for the disposal of grievances about rehabilitation of families affected by all other projects in the Narmada Valley within the state. Besides GRA, the NVDA has also setup a grievance redressal cell in the office. The resolution dealing with the constitution of GRA for projects in Madhya Pradesh provides for the appointment of two members in addition to the chairman. Later, in a petition, on the issue of rehabilitation pertaining to the Indira Sagar Project filed by the NBA, the Supreme Court has directed that in the interest of quick disposal of grievances, the GRA should be revamped to have 5 judicial and 5 administrative members. Accordingly, in the GRA, there is provision for 10 members, including the chairman.

NATIONAL REHABILITATION AND RESETTLEMENT POLICY

The unnatural and unwilling displacement of a large population continued with more projects coming into action. That too, without a uniform policy. The states had formulated their own displacement and rehabilitation policies. Whenever voices of dissatisfaction from the affected families became louder, state governments made some amendments or inducted some new provision. The major reason of discontent among the affected families was the displacement benefits decided by the government. Displacement benefits were uneven amongst the states. Serious discontent always erupted on the poor rate of compensation for an acquired agricultural land. In order to set up standard norms for the rehabilitation of people affected by various projects in the country, Government of India has notified National Rehabilitation and Resettlement Policy 2007 with effect from 31 October 2007. This policy is a modified and a fully expanded version of the National Rehabilitation and Resettlement Policy for project-affected families formulated in the year 2003. It will be imperative for the states to apply this policy to all such projects that involve non-voluntary displacement of more than 400 families in the non-tribal areas and more than 200 families in the tribal areas. It has also been stated in the policy that state governments, public sector undertakings, agencies, and other such bodies would be at liberty to put in place higher benefit levels than those prescribed in the national policy.

The salient features of the National R&R Policy 2007 are stated as follows:

- Land or house allotted to the affected families shall be in the joint names of husband and wife of the affected family.
- The affected families who have not been provided agricultural land or employment shall be paid a rehabilitation grant equivalent to 750 days of minimum agricultural wage.
- The displaced family shall get a monthly subsistence allowance equivalent to 25 days of minimum agricultural wages per month for a period of one year from the date of displacement.
- The affected persons who are disabled, destitute, orphans, widows, unmarried girls, abandoned women, or persons above 50 years of age who cannot be provided with alternative livelihood, and who are not otherwise covered as part of the affected family shall be paid Rs. 500 per month till their death.
- The affected family not availing free transport facility for shifting belongings shall be paid in lieu thereof Rs. 10,000 in one go.
- The displaced family having cattle shed shall get Rs. 15,000 for the construction of cattle shed at the new place of resettlement.
- Displaced small traders and rural artisans shall get financial assistance of Rs. 25,000 for the construction of a working shed or shop at the new place, subject to certain conditions.
- A separate Tribal Action Plan (TAP) would be formulated where 200 or more tribal families are shifted non-voluntarily, for the purpose of project construction.
- The government of Madhya Pradesh had formulated a revised rehabilitation policy in 2008 for future projects by incorporating additional beneficial provisions in the national policy.
- The rehabilitation policies of the states of Gujarat and Maharashtra also contain more liberal/beneficial provisions compared to the provisions of the national policy.

THE LAND ACQUISITION, REHABILITATION, AND RESETTLEMENT ACT 2013

The age-old Land Acquisition Act of 1894 had been subject to multiple amendments by the central and state governments. The amendments were made from time to time as the act was found to be inadequate to address important issues related to the exercise of statutory powers by the state for involuntary acquisition of private land and property, rehabilitation, and resettlement of affected families, etc.

In the developing scenario of the country, there was a need of a central law that could adequately address all the major and minor issues, which would be in the larger interest of the state in general and individuals in particular. Government of India has also formulated a new act aiming at giving the right for fare compensation and making transparency in land acquisition, rehabilitation, and resettlement.

Replacing the Act of 1894, this act came into force from 1 January 2014. The main provisions of the act are as follows:

- There would be a social impact assessment study to be conducted for every acquisition of land.
- To safeguard food security and prevent arbitrary acquisition, the act directs states to impose limits of acquisition on the area under agricultural cultivation.
- If the acquired land remains unutilized, the act empowers the states to return the land either to the owner or to the state land bank.
- If the acquired land is sold to a third party for a higher price, then 40% of the appreciated land value or profit will be shared with the original owner.
- A consent of not less than 70% of those whose land is sought to be acquired would be necessary if the land is acquired.
- For public-private partnership (PPP) projects, a consent of 80% would be required if land acquisition is taking place for a private company.

- No one shall be shifted until and unless all claims are paid and alternative sites for resettlement and rehabilitation have been prepared.
- The compensation of the land would be four times the market value in rural areas and two times the market value in urban areas.
- The act applies retrospectively to cases where no land acquisition award has been made.
- The act provides compensation to those who are dependent for their livelihood on the land being acquired.
- No income tax and stamp duty shall be charged on any amount that is accrued to an individual as a result of the provisions of this act.

The practicability of the new land acquisition act in providing land for irrigation, power, and industrial projects was questioned by a number of state governments. The political parties, too, forced to review and amend the act in the larger interest of the farming community. Ultimately, the government agreed to induct nine amendments in the act. The amended version of the bill was approved by Lok Sabha on 10 March 2015. The main amendments are:

- Land will not be acquired for private hospitals and educational institutions.
- Land acquisition will only be for industrial corridors set up by the state governments.
- Land acquired for industrial corridors cannot exceed more than 1 km on either side of the designated rail line or the road of industrial corridors.
- Compulsory employment will be made available to one member of the affected farm labourer's family in the project coming up on the land.

- Authorities to hold hearing in districts where land is being acquired.
- Social infrastructure projects are not exempted from consent clause and social impact assessment.
- Government will conduct a survey of wastelands.

However, these amendments could not be approved by Rajya Sabha within the requisite time period. Therefore, they lapsed and could not be incorporated in the act.

RESERVOIRS ENGULFING ARCHAEOLOGICAL ASSETS

The evidence obtained after survey and excavation done in the Narmada Valley reveals that the latter was an important region in the prehistoric and post-prehistoric era. Excavation in places like Maheshwar, Navdatoli, Pantnagar, Katnera, Khapakheda, etc. in Madhya Pradesh has produced evidence of copper-age culture in this region. Remains obtained in excavation at Navdatoli and Maheshwar pertain to Heheya and Yaduvanshi society. The people belonging to these clans had come here to settle down sometime in second century BC. The residents of this region have been referred to as Anup, Taljangh, Vitihotra, and Avanti in the Puranas. In this region, the chronological link of history is found since sixth century BC. Since then, the Narmada Valley is witness to the rise and fall of hundreds of rulers, royal families, marauders, satraps, kings, sultans, and empires.

In the modern era, when the development of Narmada Valley was mooted, it also included concern for protecting the invaluable historical and archaeological assets. The first phase of development of Narmada Valley consisted of the construction of dams on Narmada and its tributaries. For this reason, the protection of archaeological assets in the area of submergence becomes necessary. Hence, a working plan was formulated, mainly for the protection of archaeological monuments in Madhya Pradesh, Gujarat, and Maharashtra affected by the Sardar Sarovar Project, and in Madhya Pradesh in the region affected by Indira Sagar, Omkareshwar, and other projects.

It was stated in the NWDT award that the expenditure involved in the protection of archaeological assets by the submergence of Sardar Sarovar Project shall be borne by Gujarat. In respect of the remaining projects of Madhya Pradesh, the expenditure on these works shall be borne by the government of Madhya Pradesh. The NVDA in Madhya Pradesh has entrusted to the state department of archaeology the work of survey for identifying the archaeological monuments, which have been affected due to submergence and for its appropriate protection thereof.

The Department of Archaeology, Government of Madhya Pradesh, started the archaeological survey on a large scale in the year 1988, which was completed in the year 1992. Based on this survey, the state department of archaeology prepared a comprehensive working plan in the year 1993 for protection, study, etc. of the identified archaeological monuments and/or their remains. After holding consultations with the NVDA, this working plan was amended in the year 1997. In Madhya Pradesh, the main work of protection of the archaeological monuments in the Narmada Valley has been executed as per the working plan of 1997. The target dates for the completion of various works, as provided in the working plan of 1997, were revised in 2002.

The following works have been executed primarily for the protection of archaeological monuments:

Documentation To prepare a record of the actual status of archaeological monuments and other assets affected by submergence, and to prepare films and photographs thereof.

Reinstallation of monuments To shift ancient or historical monuments and temples from the area of submergence and to reinstall them at new places in the same form.

Excavation Excavation of important mounds in the area of submergence in order to dig out archeological assets for their protection.

Compilation of archaeological assets/monuments The collection and compilation of statues and other important remains spread over in the reservoir area and to store them at safe places.

Preparation of replica In order to preserve the original form of important monuments and statues, their replica in plaster of Paris, fibrecoat, and terracotta have been prepared.

Chemical treatment Necessary chemicals are being used on statues and monuments for their protection at the time of reinstallation.

Publication To publish photos and information about monuments affected by submergence.

Establishment of archaeological museums The work of establishing archaeological museums for safe placement and display of archaeological assets/monuments obtained from the area under submergence and from excavation.

Relocation of monuments Monuments under the reservoirs indentified, shifted, and reinstalled in their original look and position.

Archaeological monuments and their estimated time period

S. No.	Name of monuments	District/Tehsil/State	Estimated period
1	Shiv Mandir, Roligaon	Alirajpur/Alirajpur/MP	11th–12th century AD
2	Shiv Mandir, Barda	Dhar/Manavar/MP	15th–16th century AD

S. No.	Name of monuments	District/Tehsil/State	Estimated period
3	Rock Statues, Pipaldagarh	Dhar/Dharampuri/MP	12th–13th century AD
4	Shiv Mandir, Pipaldagarhi	Dhar/Dharampuri/MP	12th–13th century AD
5	Narmadeshwar Temple, Dehar	Dhar/Kukshi/MP	16th–17th century AD
6	Koteshwar Temple, Kothra	Dhar/Kukshi/MP	15th–16th century AD
7	Kanjleshwar Temple, Semalda	Dhar/Manavar/MP	15th century AD
8	Six rock caves, Khujava	Dhar/Dharampuri/MP	15th–16th century AD
9	Shiv Mandir, Chhoti Kasrawad	Barwani/Barwani/MP	18th century AD
10	Six canopies of Bhilkhera	Barwani/Barwani/MP	18th–19th century AD

Besides the reinstallation of monuments and temples, the excavation of important mounds in the area of submergence was also undertaken. The state archaeology department excavated 12 different mounds in the area of submergence of Sardar Sarovar reservoir in Madhya Pradesh.

Monuments under Sardar Sarovar Reservoir

S. No.	Place excavated	District
1	Khalghat	Dhar
2	Akalwara	Dhar
3	Katnera	Dhar
4	Kalyanpura	Dhar
5	Chhota Barda	Barwani
6	Maruchichli	Barwani
7	Kawathi	Dhar
8	Khera	Dhar
9	Brahmangaon	Dhar

Besides the excavation of the aforementioned mounds, the ASI in Nagpur also undertook an excavation of the following mounds in Madhya Pradesh:

Excavation of mounds in Nagpur, Madhya Pradesh

S. No.	Place excavated	District
1	Navda Kheri	Barwani
2	Pipri (Utavad)	Barwani
3	Khaper Kheda	Dhar

Besides Deccan College Post Graduate and Research Institute, Pune also undertook excavation at Chhoti Kasrawad in Barwani. A sum of 118 statues of archaeological importance from the areas submerged by Sardar Sarovar Project in Madhya Pradesh have been collected and kept in an archaeological museum in Dhar.

ARCHAEOLOGICAL MUSEUMS

For the safe upkeep of items of archaeological importance collected from the submergence areas of different projects of Narmada Valley in Madhya Pradesh, a big archaeological museum with seven galleries is being established at Kasrawad in Khargone. At present, the items have been displayed in three galleries. The remaining four galleries are under construction. Besides this, one archaeological gallery in the state museum of Bhopal has been established, displaying a few of such items.

Monuments under Indira Sagar Dam

S. No.	Name of monuments	District/Tehsil/State	Estimated period
1	Shiv Mandir, Dharikotla	Khandwa/Khandwa/MP	18th century AD
2	Canopy, Ghisor	Khandwa/Khandwa/MP	18th century AD
3	Shiv Mandir, Saraswati Kund	Harsud/Khandwa/MP	18th century AD

S. No.	Name of monuments	District/Tehsil/State	Estimated period
4	Rock Statues, Daiyat	Khategaon/Dewas/MP	12th–13th century AD
5	Riddheshwar Mandir, Handia	Harda/Harda/MP	12th–13th century AD

The Riddheshwar Mandir at Handia has been preserved in its original place by enclosing it with a protective wall. Due to the opposition against shifting the tomb of the famous saint of Nimar Sant Singaji, which faced the threat of submergence from Indira Sagar Dam, the tomb was reinstalled at the same place by raising it. A special approach road has been constructed to reach the tomb.

The excavation of six different mounds of archaeological importance in the area of submergence of Indira Sagar Dam was executed by Madhya Pradesh Archaeology Department.

Places and districts of the excavated mounds in the area of submergence of Indira Sagar Dam

S. No.	Excavated places	District
1	Khedinema	Hoshangabad/Harda
2	Gajanpura	Dewas
3	Bijalpur Khurd	Khandwa
4	Chapakheda	Khandwa
5	Nawalkheda	Khandwa
6	Ganaur	Khandwa

Monuments under Omkareshwar Dam

S. No.	Name of monuments	Tehsil/District/State	Estimated period
1	Shiv temple, Sailani Ghat	Khandwa/Khandwa/MP	12th–13th century AD
2	Shiv temple, Panthiya	Khandwa/Khandwa/MP	12th–13th century AD
3	Pashupatinath Temple, Panthiya	Khandwa/Khandwa/MP	18th century AD

S. No.	Name of monuments	Tehsil/District/State	Estimated period
4	Sati Mata Temple, Sailani Ghat	Khandwa/Khandwa/MP	12th–13th century AD
5	Chakkardai Temple, Sailani Ghat	Khandwa/Khandwa/MP	18th century AD
6	Shiv temple, Premgarh	Bagli/Dewas/MP	18th century AD

Besides the aforementioned ones, the 24 avatar (incarnation) temples dedicated to the 24 forms of Lord Vishnu, affected by Omkareshwar Project, was shifted and reinstalled by ASI. This temple is included in the list of protected monuments of ASI.

Monuments under Maheshwar Dam

Madhya Pradesh Archaeology Department had surveyed and identified the monuments and archaeological remains spread over 61 villages affected by Maheshwar Dam poundage on river Narmada.

Monuments under Maheshwar Dam

S. No.	Name of monuments	Tehsil/district/state	Estimated period
1	Vimleshwar Temple, Belsar	Barwah/Khargone/MP	16th–17th century AD
2	Kaleshwar Temple, Mardana	Barwah/Khargone/MP	17th–18th century AD
3	Peshwa Bajirao Tomb, Raverkhedi	Maheshwar/Khargone/MP	18th century AD

The Maheshwar Project is being constructed by the Energy Department of Madhya Pradesh through the private construction agency Shree Maheshwar Hydel Power Co. (SMHPC). It is for this agency to organize the protection work of the monuments affected by submergence through the state archaeology department. The Peshwa Bajirao tomb at Raverkhedi is a classified monument protected by the

ASI. The reinstallation of the remaining monuments remains to be done by the state archaeology department.

The mound at Patnagar in the tehsil of Barwah in Khargone has been identified for excavation during the archaeological survey of the affected area. During the survey, the remains of utensils belonging to the Copper Age and bangles of couch were recovered. An extensive excavation in the area has been proposed.

Monuments under Rani Avanti Bai Sagar Dam

The archaeologically important temple Nandikeshwar Mahadev, situated at village Bhidki, affected by the submergence caused due to the project reservoir has been protected. This temple has been reinstalled at Bargi Nagar. The Nandikeshwar temple finds mention in the Puranas. The importance of this temple has been described in *Matsya Purana*, *Shiva Purana*, *Skanda Purana*, *Markandeya Purana*, and *Linga Purana*.

Monuments in Gujarat and Maharashtra

The State Archaeology Department of Gujarat had conducted a survey of the area affected by Sardar Sarovar Dam in Gujarat and Maharashtra. It was found that none of the protected monuments is affected by submergence due to the reservoir. According to the survey report, Surpaneshwar Shiva temple on the border of Gujarat and Maharashtra was impacted, where the villagers assemble for pilgrimage each year. Similarly, Hampheswar Mahadev temple situated in Chhota Udaipur taluka of Panchmahal in Gujarat is also affected, where people go for pilgrimage. Both these temples are not protected monuments. Looking at the religious importance, Surpaneshwar Shiva temple has been shifted to a place called Gora near Kevadiya. The Hampheswar Mahadev temple has also been reinstalled at a safe place.

FOSSILS FOUND IN NARMADA VALLEY

Large numbers of fossils have been discovered at various places in Narmada Valley. These fossils may provide deep insights into the history of the valley and could unravel several myths about evolution and early civilization. The most significant and rarest fossil discovered in the valley is the intertrappean fossil. It is the only human fossil discovered in Asia. The place where the fossil was discovered is 10 kms away from the village Jait in Sehore district of Madhya Pradesh. A proper study of the intertrappean fossil can throw much light on the early life on earth. The fossil was discovered in the year 1982 by Arun Sonakia, former director of Geological Survey of India and a paleontologist. According to him, the human fossil found was approximately five hundred thousand years old and belonged to the homo erectus species, the predecessor of today's human beings.

The fossils of starfish, shark, and many other marine lives have also been found in Dhar region of Madhya Pradesh. This suggests that the area was under the sea. Fossils of eucalyptus plants have also been discovered, which are usually found in Australia. Traces of the last walk of the dinosaurs were found at Mandu (Dhar) near Rani Roopmati Mahal. The region has dinosaur fossils from three different generations, the fossils of which are very significant to decode in order to understand the reason behind their extinction and finding out how long they survived on earth.

The discovery of around 86 million-year-old fossil of a rare species of an oyster has once again reconfirmed the presence of marine life and sea in Central India. This fossil was recently discovered in Dhar, Madhya Pradesh by a paleontologist. The fossil is identified as belonging to the crassostrea species considered to be one of the oldest evidences of sea incursion in the Narmada Valley. A large number of marine molluscs like oysters and gastropods are found to be the matrix of this fossil.

It is one of the best-quality fossil sample found in the sandstone of Bagh Beds in Dhar. The sediment of this sea is defined as the upper cretaceous age, which is from 140 to 65 million years ago. In the year 2012, fossilized teeth and bones of three generations of sharks were discovered at Bagh belt in Dhar district. These are believed to be 6.5 crore years to more than 10 crore years old.

Fossils found in Narmada Valley can greatly help in understanding the migration route of humans, evaluation of life on earth, and unravelling many more untold mysteries. The extensive research on fossils found in Narmada Valley may also put several existing notions to rest.

Fossil of Narmada man—*Homo erectus narmadensis*

Some of the fossils discovered in the Narmada Valley till date are:

- Narmada man (Homo erectus narmadensis)
- ostrich (Struthio camelus)
- eucalyptus plant
- carnivore (Holarctus namadicus)
- lion (Panthera leo)
- panther (Panthera pardus)
- sloth bear (Urus namadicus)
- wolf (Canis aureus)
- stegodon (Stegodon namadicus and Stegodon insignis)
- primeval elephant (Elephas namadicus and Elephas maximus)
- wild dog (Cuon alpinus)
- horse (Equus namadicus)
- Asiatic wild horse (Equus hemionus khur)
- rhinoceros (Rhinocerous unicornis)
- hippopotamus (hexaprotodon namadicus)
- guar (Bos gaurus)
- buffalo (Bos primigenius namadicus)
- blue bull (Boselaphus namadicus)
- black buck (Antilope cervicapra)
- sambhar or swamp deer (Cervus unicolour)
- chinkara (Gazella bennettii)
- barasingha/swamp deer (Cervus duvaceli)
- spotted deer (Axis axis)
- wild boar (Sus namadicus)
- porcupine (Hystrix crassidens)
- primordial lizard (Agama schleichi)
- tortoise (Trionyx)
- crocodile (Crocodylus palaeindicus)
- marine plants, sharks, and other marine beings

The fossils discovered in Narmada Valley are displayed at different museums, mainly in Madhya Pradesh:

- Narmada Fossil Museum, Indore (MP)
- State Museum, Bhopal (MP)
- Ashmedha Fossil Museum, Mandla (MP)
- Archeological Museum, Mandla (MP)
- Ghughwa Museum, Dindori (MP)
- Kheda Museum (Gujarat)

FOREST, WILDLIFE, AND ENVIRONMENT

The entire Narmada Valley brims with forest wealth. The forests in the valley, full of innumerable species of flora, are spread out in the Vindhya ranges on the northern bank and in Satpura ranges on the southern bank of Narmada. These forests of Vindhya and Satpura ranges are like veins and nerves of Narmada and its tributaries. The rainwater absorbed by the forests ultimately reaches Narmada in the form of small streams and rivulets.

The forests in the Narmada Valley can be divided, based on the main species of the forest, into three classes:

- Sal (Shorea robusta) forest
- Teak (Tectona grandis) forest
- Mixed/miscellaneous forest

The areas which abound in trees of sal (Shorea robusta) are put in the category of sal forest. This class of forests constitutes about 10–12% of the total forests in Narmada Valley. The forest areas which abound in teak (Tectona grandis) are classified as teak forest. They constitute nearly 30–40% of the entire forest area. Besides two broad classifications, there exist forests where sal and teak species are comparatively fewer in number and other miscellaneous species exist in excess. Such forests are classified as miscellaneous forests. The miscellaneous forests cover nearly 50% of the entire forest area in the valley.

Since a few decades, the human pressure on the forests in Narmada Valley is constantly on the increase. The population here has increased five times between 1901 and 2001. Now, the decadal growth of population is nearly 25%. Hence, one can imagine the pressure of human population on forests in recent years and in the coming ones. Where the increasing pressure of population has affected forests, there is a need for development projects for the sake of the burgeoning population. In the context of Narmada Valley, the irrigation projects cause an adverse effect by submerging large tracts of forests.

The Sardar Sarovar Project in Gujarat and various projects in Madhya Pradesh submerge vast forest areas in Gujarat, Madhya Pradesh, and Maharashtra. The submergence of forest areas, in turn, affects wildlife. In the decades following 1980, the Ministry of Environment and Forests had formulated a fairly detailed enquiry for conducting evaluation studies of the environmental impact of Narmada Valley projects and formulating plans for conservation of the environment. Based on this, the states of Madhya Pradesh, Gujarat, and Maharashtra has submitted working plans pertaining to forests and environment. After studying the working plans and holding discussion at various levels, on 24 June 1987, the Ministry of Environment and Forests accorded environmental clearance to Sardar Sarovar Project in Gujarat and Indira Sagar Project in Madhya Pradesh, subject to certain conditionalities.

The other projects in Narmada Valley in Madhya Pradesh were also accorded sanction by the ministry only after compliance with the necessary conditions regarding the conservation of forest and environment.

The condition imposed in the sanction of Narmada Valley projects by the Ministry of Environment and Forests is that in lieu of every hectare of forestland affected either by reservoir submergence or by the construction of a project, a hectare of non-forestland should be afforested. If compensative afforestation is done on degraded forestland, afforestation in two hectares would be required for every hectare of forest affected. In the same way, cutting of trees in forests affected by submergence and transportation thereof would be ensured by the concerned state in the interest of the environmental safety of the reservoir.

A total of 13,385 hectares of forestland in the states of Gujarat, Madhya Pradesh, and Maharashtra has been affected by the submergence caused by Sardar Sarovar Project. As per directions of the Ministry of Environment and Forests for creating compensatory forests in lieu of the submerged forestland, plantation work was undertaken. In the three states, in lieu of 13,385 hectares of submerged forestland, the plantation of forest species was done on 42,066 hectares. These newly created forest areas in the three states were notified as forest areas under the provisions of The Indian Forest Act 1927.

FORESTS SUBMERGED IN GUJARAT

In lieu of 4,166 hectares of forestland affected by the submergence of Sardar Sarovar Project in Kachchh, Gujarat, 4,650 hectares of non-forestland and 9,300 hectares in nearby districts have been brought under compensatory afforestation. Thus, in Gujarat, as against 4,166 hectares of forestland affected by the submergence of Sardar Sarovar Project, 13,950 hectares of land was brought under new afforestation right before filling the reservoir. Besides, for ensuring greenery around Sardar Sarovar Dam, and as compensation to the environment, afforestation has also been done on 551 hectares near the dam site.

Sardar Sarovar Narmada Nigam Ltd has also formulated an ambitious programme of plantation along the project canal system. Plantation has already been done on 4,000 hectares of adjoining canal system. Further extending the compensatory afforestation work, 200 hectares of ravine land of Firozpur and Ratanpur on the bank of river Sabarmati near Gandhinagar has also been brought under afforestation. The available land in the residential colonies in the state has also been brought under plantation.

FORESTS SUBMERGED IN MAHARASHTRA

In lieu of 6,488 hectares of forestland affected by the submergence owing to the Sardar Sarovar Project, 19,378 hectares of land has been brought under new afforestation. Besides, in lieu of forestland used for rehabilitation work, 4,198 hectares of additional land has been afforested.

FORESTS SUBMERGED IN MADHYA PRADESH

In Madhya Pradesh, in lieu of 2,731 hectares of forestland affected by the submergence by Sardar Sarovar reservoir, 8,737 hectares of land has been brought under afforestation.

Statewise compensatory afforestation (in hectares) under Sardar Sarovar Project

State	Submerged forest area	Afforestation work done	Total afforested area
Madhya Pradesh	2,731.00	In degraded forest – 6,538 Non forest area – 2,199	8,737
Gujarat	4,165.91	In degraded forest – 9,300 Non forest area – 4,650	13,950
Maharashtra	6,488.54	In degraded forest – 12,977 Non forest area – 6,402	19,379
Total	13,385.45		42,066

SUBMERGENCE OF FORESTS UNDER OTHER DAM PROJECTS

As per the condition imposed by the Ministry of Environment and Forests while according the clearance for dam projects, compensatory forests have been created in lieu of forests affected by the submergence. Compensatory afforestation has been done by the NVDA and the state forest department.

Submergence of forests affected by the major projects in Madhya Pradesh and compensatory afforestation (in hectares)

Name of project	Affected forest area	Afforested area
Indira Sagar	41,460.930	81,453
Omkareshwar	5,829.850	11,600
Shaheed Chandra Shekhar Azad (Jobat)	130.810	385.577
Maan	4.000	4.000
Upper Beda	14.240	51.240
Lower Goi	1.083	1.083

A total of 8,478 hectares of forestland was affected by the submergence of Rani Avanti Bai Sagar Project. The forest and environment clearance of this project was obtained before the year 1980. Since the Forest Conservation Act was not in force prior to 1980, compensatory afforestation in lieu of forestlands submerged was not compulsory.

Besides dam reservoirs, the construction of canals has also damaged forest areas. The construction of canals of Indira Sagar, Omkareshwar, and Upper Beda projects have respectively caused damage to 331.978 hectares, 17.053 hectares, and 36.727 hectares of forestland. In lieu of these forest areas, equivalent areas elsewhere have been afforested. The Bargi Diversion Project is entirely a canal project. A total of 685.286 hectares of forestland has been used for this. The forest department has created new forests on as much area.

WILDLIFE

Before according sanction to projects, the Ministry of Environment and Forests also ensures the conservation of wildlife affected by the submergence in the concerned states. For conservation of wildlife in Gujarat, Madhya Pradesh, and Maharashtra, studies have been conducted by expert agencies regarding environmental impact on wildlife. Based on the conclusions of these studies, the three states have taken steps for the conservation of wildlife.

WILDLIFE AFFECTED BY SARDAR SAROVAR RESERVOIR

For the conservation of wildlife affected by the Sardar Sarovar Project in Gujarat, the area of preexisting Dumkal Bear Sanctuary has been extended and developed with the new name of Shoolpaneshwar Wildlife Sanctuary. The initial 102 sq. km of this sanctuary has been extended to 608 sq. km, and necessary development work has been done for animals. The wild animals subsisting on the left bank of Sardar Sarovar have been provided shelter in Shoolpaneshwar Sanctuary. In the Rann of Kutch, an initiative has been taken for conserving and developing the great Indian bustard.

Studies have been conducted about the environmental impacts in the Nal Sarovar Bird Sanctuary, Wild Ass Sanctuary, and Blackbuck National Park, Velavadar, situated in the command area of the reservoir. Based on these studies, the working plan for the development of Nal Sarovar Bird Sanctuary and Blackbuck National Park has been executed.

In a study carried out in Maharashtra, it was recommended that to provide safety to wild animals, a buffer zone should be created. Acting on this recommendation, 25 lakh plants of local species have been cultivated on 2,500 hectares. Local varieties of plants in future would let forests grow in a natural way, resulting in the conservation of wild animals.

Environmental studies done in Madhya Pradesh have recommended that in the context of Sardar Sarovar Project, Katthiwada Wildlife Sanctuary spread over 110 sq. km in the district of Jhabua should be developed. Following the notification by the state forest department, the development work of Katthiwada Sanctuary would be executed.

WILDLIFE AFFECTED BY INDIRA SAGAR RESERVOIR

Indira Sagar reservoir affects 41,460.930 hectares of forest area. Traditionally, this area has been a safe haven for wild animals. Facing the threat of submergence, it has become necessary to conserve them. The government of Madhya Pradesh got this area surveyed and studied before submergence by an organization called Friends of Nature, working in the field of wildlife conservation. The study report revealed that the submerged area was a habitat for nearly 2,000 wild animals of 32 species. The species which were taken in to account are nilgai (blue bull), chinkara (spotted deer), hyena, tiger, blackbuck, fox, etc.

The Ministry of Environment and Forests issued directions for making arrangements for the conservation of the affected wildlife before filling the reservoir. In compliance to these directions, the state government has constituted a committee to give suggestions on these issues. The committee has recommended the establishment of a national park and sanctuary for the wildlife affected by Indira Sagar reservoir.

Accordingly, 293.56 sq. km area for Omkareshwar National Park, 178.21 sq. km for Singhji Sanctuary, and 69.24 sq. km for Mandhata Sanctuary have been identified in district Khandwa. Besides, areas of 135.03 sq. km and 20.87 sq. km have also been marked for establishing two reserve units for wild animals. Since the identified areas of national parks adjoin Indira Sagar reservoir, the possibility of pisciculture in the reservoir gets affected due to which the official notification of the national park has been kept on hold. However, the NVDA has carried out various developmental works for making the identified areas hospitable for wild animals.

AQUATIC LIFE AND PISCICULTURE

In the award of the NWDT, the states of Madhya Pradesh, Gujarat, and Maharashtra have been authorized to undertake the conservation of aquatic life and pisciculture in the reservoir of Sardar Sarovar falling within their territory. At a FRL of 138.68 metres, the water area of Sardar Sarovar reservoir suitable for pisciculture extend to nearly 37,533 hectares. Of this, 7,112 hectares, 20,822 hectares, and 9,599 hectares fall respectively within the territory of Gujarat, Madhya Pradesh, and Maharashtra. Thus, the scope of pisciculture in these states has increased.

According to an estimate, the water area of nearly 10,000 hectares in Madhya Pradesh, 3,500 hectares in Gujarat, and 3,000 hectares in Maharashtra would be assuredly available for pisciculture and production. In order to exploit the possibilities of these two practices, the three states are sowing seedlings of appropriate species in their part of the reservoir. By now, Gujarat has sown 488 lakh fish seedlings in the reservoir. Maharashtra and Madhya Pradesh have respectively sown 13.5 lakh and 4 lakh fish seedlings in their respective territories.

Every year, these states are collecting fish seedlings. In order to restrict the growth of water grass in Narmada canal, Gujarat has collected 7 lakh fish of the species that feed on this grass. There are possibilities of producing fish in Gujarat and Maharashtra every year worth Rs. 15 crore and Rs. 10 crore respectively. Similarly, Madhya Pradesh can produce fish worth about Rs. 25 crore every year from Sardar Sarovar reservoir.

The reservoir of other projects constructed in Narmada Valley also bristle with the possibility of pisciculture and production. The area of Indira Sagar reservoir in Madhya Pradesh extends to 913 sq. km. Of this area, 59,000 hectares of water area lends itself for fish production. Water area of 8,330 hectares in Omkareshwar Dam, 696 hectares in Maan Dam, 946 hectares in Jobat reservoir, and 830 hectares in Upper Beda reservoir is abundant with possibilities of fish production.

Central Inland Fisheries Research Institute has conducted studies on Indira Sagar Dam and made certain recommendations for fish production. Appropriate infrastructure is needed for exploiting fish production in its full potential in the reservoirs of Sardar Sarovar and other projects in Narmada Valley. With the help of appropriate infrastructure, besides commercial fish production, the possibilities of employment can also be exploited. It is estimated that Narmada Valley reservoirs in Madhya Pradesh can yield fish worth Rs. 50 crore every year.

RESERVOIRS AND CATCHMENT AREA

The geographical area which feeds rivers and reservoirs is known as the catchment area. Catchment area is the area over which water flows into the river/reservoir. This flow of water is collected directly by tributaries and rivulets, and indirectly by the trickling down of rainwater falling on ground. The water flowing from the catchment area also brings soil with it. Thus, after an interval, silt gets deposited in the reservoir. With continuous deposition of silt, the water-bearing capacity of the reservoir goes on diminishing. This could lead to considerable reduction in the water-holding capacity of the reservoir over a long course of time. As a solution, the catchment area is scientifically treated in a manner that water flows over carrying minimum soil.

For ensuring a longer life of the reservoirs, it is extremely essential that an appropriate treatment is carried out in the catchment areas of major reservoirs. The catchments of the reservoir comprise both forest areas and non-forest areas. In the forest areas, the catchment area treatment work include check dam, contour trenching, plantation, seed-sowing, development of grazing land, loose boulder, earthen bund, grabian tank structure, percolation tank, water harvesting structure, etc. In the non-forest areas, the treatment work includes construction of earthen dam, boulder check dam, (boulder) plantation of appropriate flora, plantation of fruit-bearing trees, straightening course of streams of rainwater, controlling speed of waterflow, etc.

The total catchment area of Sardar Sarovar in the three states is 24,42,441 hectares. This area has been divided into two categories. The first category comprises of the straight flow of water into the reservoir from rivers and rivulets; the second category includes free-flowing water area. In Madhya Pradesh, the catchment area treatment work has been done on 1,61,305 hectares. Of this, an area of 1,10,997 hectares is straight flow area, whereas 50,308 hectares pertain to free flow area. In Gujarat, treatment work has been done on 29,657 hectares, of which 29,157 hectares belong to straight flow category and the remaining 500 hectares is free flow area. Similarly, in Maharashtra, treatment has been done on an area of 44,954 hectares, of which 23,295 hectares pertain to straight flow and 21,659 hectares fall under the free flow. The treated area comprises both forest and non-forest areas.

The catchment area treatment has also been done in other major projects in Narmada Valley. Of the total area of 61,352 hectares of Indira Sagar Project, treatment work has been executed on a treatable area of 55,523 hectares. In Omkareshwar, out of 28,455 hectares of area, treatment has been executed on 24,514 hectares. Treatment work is still going on in the catchment areas of various projects.

The position of catchment area treatment in various projects (area in hectares)

Name of project	Category of area	Catchment area	Treated area		Total treated area
Indira Sagar	Straight water flow	61,352	Forest area	5,686	55,523
			Non-forest area	49,837	
	Free water flow	9,15,150	Forest area	18,378	30,513
			Non-forest area	12,135	
Omkareshwar	Straight water flow	28,455	Forest area	10,851	24,514
			Non-forest area	13,663	
	Free water flow	50,833	Forest area	—	Catchment Area Treatment (CAT) to be done
			Non-forest area	—	

Name of project	**Category of area**	**Catchment area**	**Treated area**		**Total treated area**
Maan	Straight water flow	5,148	Forest area	5,148	5,148
			Non-forest area	0	
	Free water flow	—	Forest area	—	—
			Non-forest area	—	
Jobat	Straight water flow	16,586	Forest area	386	16,485
			Non-forest area	16,099	
	Free water flow	33,141	Forest area	—	CAT to be done
			Non-forest area	—	
Upper Beda	Straight water flow	1,855	Forest area	181	1,850
			Non-forest area	1,669	
	Free water flow	6,234	Forest area	—	CAT to be done
			Non-forest area	—	
Maheshwar	Straight water flow	15,614	Forest area	—	11,626
			Non-forest area	11,626	
	Free water flow		Forest area	—	—
			Non-forest area	—	

Command area refers to the area which comes under irrigation in the project. An optimum utilization of created irrigation potential depends on an equitable distribution of water and proper development of the command area. Development plans have been prepared for the command area of Sardar Sarovar Project in Gujarat and Rajasthan, and for the command areas of Indira Sagar, Omkareshwar, and other projects in Madhya Pradesh.

Command area development work includes levelling of lands, making various formations/channels in the fields for flow of water, training farmers in latest scientific, agricultural practices, and making improved seeds and tools available to them. This ensures an effective command of a major irrigation project.

For command area development, the assistance of non-government organizations (NGOs) has been sought in Gujarat, Madhya Pradesh, and Rajasthan. These states have also introduced water management system with the active participation of farmers. Under the participatory water management system in Madhya Pradesh, para-social workers (*kolaba sahayaks*) have been appointed. Water-utilization societies of farmers have selected 7,170 para workers. These workers are being trained in irrigation, water management, *barabandi*, etc. In Madhya Pradesh, in the command areas of Maan and Jobat (Shaheed Chandra Shekhar Azad) projects, 100% utilization of created irrigation potential is taking place through participatory irrigation management.

Participatory irrigation management is the only way to utilize maximum created irrigation potential of an irrigation project. But the effective implementation of this system for the right water governance in irrigation projects in Narmada Valley is still to be done seriously. Otherwise, the absence of water governance may cause serious consequences, which may hamper the ultimate objective of irrigation projects.

RESERVOIRS AND HUMAN HEALTH

Large water bodies, with their stagnant water, create ideal conditions for water- and vector-borne diseases to flourish. This aspect has been a matter of major concern associated with dam projects. In the context of Sardar Sarovar Project, the National Rural Health Mission (NRHM) got some basic studies conducted in Gujarat, Madhya Pradesh, and Maharashtra. All the three states have formulated health working plans on the healthcare requirements of Sardar Sarovar. In the reservoir areas of the three states under the NRHM, regular surveys and investigation work are undertaken. In Kevadiya (Gujarat), a 50-bedded hospital has been established. Within the periphery of the reservoir healthcare laboratory, 20 mobile hospital units and 61 dispensaries in various villages have been extending healthcare services.

In Madhya Pradesh, 1 hospital, 1 small primary health centre, and 58 dispensaries have been established within the periphery of the reservoir. In Nisarpur, near the reservoir area, a 30-bedded hospital has been established and handed over to the state health department for running it. Similarly, the hospital in Dharampuri falling within the periphery of the reservoir has been upgraded. In Maharashtra, 4 rural hospitals and 26 primary health centres have been established. Also, 2 floating dispensaries, 111 health sub-centres, and 25 rescue centres within the periphery of the reservoir have been opened.

Healthcare working plans have also been prepared for Indira Sagar, Omkareshwar, Rani Avanti Bai Sagar, Maan, Jobat, and Upper Beda projects in Madhya Pradesh. The hospitals, in the districts within the affected area of these projects, have been specially upgraded for malaria and other waterborne diseases.

Madhya Pradesh has availed the services of the National Institute of Malaria Research (NIMR) for studying possibilities of the likely diseases in the habitations in the periphery of the reservoirs, and for taking necessary prophylactic measures for prevention thereof.

With a memorandum of understanding (MoU) signed for this, the experts from NIMR, New Delhi, visited the reservoir areas of the projects and undertook necessary survey, study, and research work. The NIMR has established three healthcare units in Bhopal, Jabalpur, and Narmada Nagar near Khandwa in Madhya Pradesh.

These units make regular inspections in the villages of the reservoir and command areas of the project and collect blood samples. The blood samples are sent to NIMR, New Delhi Laboratory, for detecting malaria and other diseases. Healthcare units also extend medical advice and provide medicines.

NAVIGATION IN NARMADA

The possibility of interstate water navigation in Narmada finds mention for the first time in the report of an ad hoc committee on transportation matters constituted by the government in 1948. In this report, mention had been made about the possibility of transport by water in Narmada from Hoshangabad to the point where it meets the Arabian Sea. Since the subject of navigation comes within the jurisdiction of Government of India, an interstate water transport committee was constituted. Later, in 1959, the committee mentioned in its report that a major part of Narmada till Bharuch (Gujarat) was suitable for water transport. In deep waters, transportation is made possible by ships weighing 70 to 80 tonnes. In shallower areas, transportation is possible by small boats.

The committee had made a few recommendations for making Narmada conveniently suitable for water transport. It further observed that for developing master plans for inland water transportation, the river Ganga should be linked with westward-flowing rivers. For developing the waterway from Arabian Sea to the Bay of Bengal (west to east), Narmada should be linked to Sone, Chambal, Yamuna, and Godavari rivers.

No action was taken on the suggestions of the committee. However, in the year 1960, Government of India sent the report to the state governments and asked them to keep in mind, while constructing dam projects on various rivers, that resources for multipurpose development would be available in the river basin. It should also be ensured that the construction of reservoirs is done in a well-organized way rather than in a haphazard manner, so that the rivers and reservoirs lend themselves for water transportation.

Besides the construction of various dam projects on Narmada, for assessing the utility of water transportation from Hoshangabad to Bharuch, Government of India got the quality of water surveyed by a team of experts from Netherlands in 1981–1985. This team, besides giving various suggestions, also elaborated on the technique to enable ships cross the dams between Hoshangabad and Bharuch. The suggested technique is being successfully used by a few European countries.

Government of India directed the Water and Power Consultancy Services (WAPCOS) to examine the suggestions made by the team of experts from Netherlands. WAPCOS, without recommending any technique for crossing the dams, suggested that the possibilities of transport within the reservoir should be exploited. The governments of Madhya Pradesh and Gujarat, while stating that ship transport between Hoshangabad and Bharuch should be economically viable, have asked the government to take initiative on this behalf. Five national waterways have been notified in the country till now. So far, no firm initiative has been taken for making Narmada basin a national waterway.

THE RIVER OF PILGRIMAGE

Since time immemorial, pilgrimages have been made on the banks of Narmada. Every year, thousands of people perform Narmada parikrama by walking from the origin of Narmada to its end on both sides of the bank. There are large numbers of temples, especially Shiva temples, located on both banks of Narmada. The most important among them is jyotirlinga of Omkareshwar in Madhya Pradesh, which is one of the 12 jyotirlingas in India.

Narmada Valley reservoirs have created obstructions in the age-old Narmada parikrama because many temples and traditionally used pathways have come under submergence. The temples affected by reservoirs have been shifted and reinstalled to safer places in Madhya Pradesh and Gujarat. To provide comfort to pilgrims in performing the Narmada parikrama, Madhya Pradesh has come up with a scheme to relocate and upgrade the pathway for the purpose. Under this scheme, not only an alternative footpath has been constructed in place of the submerged one, but night shelters and other basic facilities on this route is also provided to the pilgrims.

Before the construction of projects on Narmada, the total length of circumambulatory route on both sides of the bank, from the origin to the end, was 2,721 kms. From the origin to the border of Madhya Pradesh, the length of this route was 2,073 kms. Because of reservoirs of Sardar Sarovar, Indira Sagar, Omkareshwar, Rani Avanti Bai Sagar, and Maheshwar dams, the length of the circumambulatory route in Madhya Pradesh has increased by 25%, and would now be about 2,600 kms. Of this 2,600 km long footpath, 1,560 kms would be developed anew. The remaining 1,040 kms of old route shall be upgraded as per new requirements.

The survey and marking of this route is being done by the NVDA. The responsibility of executing the remaining work of the route has been entrusted to the concerned departments. According to the working plan, the work relating to the construction of Narmada circumambulatory route for pilgrims started in 2009–10, and was estimated to be completed within a period of five years.

In Gujarat, on both sides of Narmada, there are many spots of religious importance. Shurpaneshwar temple, affected by the submergence of Sardar Sarovar reservoir, has been reinstalled near Kevadiya. Another temple Hampeshwar has also been reinstalled. Amongst the prominent places of Narmada pilgrimage in Gujarat, Chandod, situated at a distance of 55 kms from Vadodara at the confluence of the rivers Narmada, Orsang, and Saraswati, is an important religious place. On the bank of Narmada, near Vadodara-Bharuch National Highway No. 8, the place of Narmada pilgrimage Nareshwar is revered as a holy site. Kabirbad is a religious place near Bharuch. Near Rajpipla, the headquarters of district Narmada, Garudeshwar is a sacred place on the banks of Narmada, where the temples of Narad, Garud, and Koteshwar, and the tomb of Swami Vasudevanand Saraswati Maharaj are located.

There are prospects of tourism in the modern centres of development in the Narmada Valley. Big dams, reservoirs, and greenery spread across reservoirs and mountain ranges attract tourists. Gujarat has developed Sardar Sarovar Project station as a tourist centre. The tourists are charged a nominal fee and allowed to roam. Since the construction of the project, lakhs of tourists have visited this place. In Madhya Pradesh, Indira Sagar, Omkareshwar, and Rani Avanti Bai Sagar project stations can be developed as good tourist spots. Many big islands inside reservoirs bristle with imminent possibilities of tourism.

NARMADA AMID LEGAL HURDLES

The major Narmada Valley projects have always been subject to judicial scrutiny. NBA, the leading NGO, has been ahead in filling petitions on behalf of the project-affected persons before the high court and the Supreme Court. All these petitions have been based mainly on the issues of environment, rehabilitation of the affected families, compensation, and other reliefs.

In the petitions filed with regard to various projects, citing incomplete and unsatisfactory rehabilitation as the main basis, demands like staying the work on the project, prohibiting the filling of reservoirs, and providing more relief to the displaced families have generally been raised.

A writ petition was filed before the high court in Jabalpur in 2007 by the NBA, alleging corruption in Madhya Pradesh in the disbursement of special rehabilitation package for Sardar Sarovar Project. Under the package, it was imperative to produce purchase agreement with the seller after which the second installment of the package could be released for the purchase of land. It was alleged by the NBA that the middlemen and officers colluded to create fake registration documents of the land purchase, made the displaced farmers submit them, and indulged in corruption while paying the package amount to them. In the writ petition, directions have been sought to register criminal cases against officers and middlemen at fault, to engage the CBI to investigate the matter, and to ask the Comptroller and Auditor General to get the audit done.

The high court in Jabalpur, while passing the interim order on the case on 22 August 2008, directed the government of Madhya Pradesh to constitute an enquiry commission. In compliance of the order, Madhya Pradesh constituted an enquiry commission under the chairmanship of Justice S.S. Jha. The commission had to probe the matter. The issue of filling of the reservoirs of Indira Sagar and Omkareshwar projects upto FRL were also put many times before the courts by the NBA.

In 2009, the NBA filed a petition in Madhya Pradesh High Court, seeking directions for staying the acquisition of land for the construction of Indira Sagar and Omkareshwar projects on the ground that the state government did not hold prior consultation with gram sabhas before starting land acquisition, as required under the provision of the Panchayats (Extension to Scheduled Areas) Act 1996 (PESA). The high court, by its interim order dated 1 July 2009, stayed the excavation of canals and the acquisition of land for canal, holding that the provisions of the PESA Act had not been complied with.

On 11 November 2009, in the final order, the high court held that the PESA Act was not applicable in the case. However, the high court continued the stay for the reason that the command area development plan had not been prepared. It was also ordered by the high court that the persons affected by the acquisition of land for canals be extended benefits under the rehabilitation policy.

Madhya Pradesh filed a Special Leave Petition (SLP) for canals before the Supreme Court. The Supreme Court, by its interim order dated 25 February 2010, has vacated the stay imposed on the acquisition of land for the canals, and directed that those cases where more than 60% landholding is acquired from a person for the construction of canals should be treated as hard cases, and the benefit of rehabilitation package should be extended to them by the state government.

In the petition filed by the NBA on the issue of rehabilitation of five villages affected by the Omkareshwar Project, the high court passed orders on 23 September 2009 stating that the benefits of rehabilitation be extended to the families of these villages. Madhya Pradesh challenged this order in appeal before the Supreme Court saying that no land in these villages had been acquired. However, the NBA in their affidavit maintained that the acquisition of land took place in these villages. The Supreme Court got the facts mentioned in the affidavit verified by the district judge in Indore. It was found that the land was not actually acquired by the government. On 5 April 2011, the Supreme Court issued a notice to the NBA, asking why it should not be blacklisted for misleading the court by providing false information in the affidavit. The Supreme Court, disposing of the appeals filed by both sides, passed its order on 11 May 2011 that the provision of the award by the NWDT do not apply to other projects, including Omkareshwar.

A view of public protest against Omkareshwar Dam organized by the NBA

The Supreme Court observed that the courts in the country should be very cautious while perusing the petitions of this organization. However, the apex court has refrained from starting contempt proceedings against this organization.

By citing issues of rehabilitation, the NBA filed a petition in Madhya Pradesh High Court in 2008 pertaining to the Upper Beda Project. Directions to not fill the reservoir were sought on the ground that rehabilitation was not completed. The high court later disposed the petition, directing the state government that the reservoir may be filled up to 310 metres, and that after completing the canal network, a fresh application should be submitted before the high court for granting permission to fill the reservoir up to 317 FRL.

In the context of directions by the court, the NBA has from time to time filed contempt petitions. Many of such petitions are pending in the high court. A petition on Maheshwar Hydroelectric Project is also pending before the high court.

A fresh petition has been filed in the high court by the NBA against the installation of gates in Sardar Sarovar Dam. On the request of the state, the high court transferred the case to the Supreme Court. The permission to install gates on the crest level of the dam was accorded by the NCA on 12 June 2014.

It is presumed that the future Narmada Valley projects would also have to face legal hurdles as part of their struggle to acheive the final goal.

WATER UTILIZATION VIS-À-VIS TIMELINE

The NWDT had declared the award with the proviso of utilizing allocated water within 45 years from the declaration of the award. The award was declared in the year 1979. Thus, all the states sharing Narmada water has to ensure the utilization of water by the year 2024. The award states that after the said timeline, the allocation of water may be reviewed and unutilized water may be reallocated to the states who deserve to utilize it.

There is a point of concern with the states—if they fail to honour the timeline for the utilization of water, they may lose unutilized water. On the basis of the quantity of allocated water and projects proposed to ensure utilization, the tasks of the states are not equal. Gujarat has to create 18.62 lakh hectare irrigation potential by utilizing the allocated 9 MAF of Narmada water. For this, Gujarat had to construct only one dam irrigation project (Sardar Sarovar) with 458.318 km main canal and distribution network. As of now, Gujarat has completed the dam and the main canal. It is likely to complete the whole target before the estimated timeline. Rajasthan has to construct 74 km long irrigation canal in extension to the Sardar Sarovar main canal from the border of Gujarat. Rajasthan has completed this canal to create 2.46 lakh hectare irrigation potential, utilizing 0.50 MAF of allocated water. Maharashtra has to utilize only 0.25 MAF of allocated water.

The most difficult task is for Madhya Pradesh to utilize 18.25 MAF of allocated water within the timeline. For this, Madhya

Pradesh has to bring 27.50 lakh hectare agricultural land under irrigation through 29 major, 135 medium, and nearly 3,000 small dams. Madhya Pradesh has had to bear maximum adversities in bringing up the major dam projects. The most challenging task has been the displacement and rehabilitation of tens of thousands of rural families from the area of future reservoirs. The dissatisfaction over the rehabilitation and strong opposition from the people has been putting hurdles time and again. Many NGOs and environmentalists, specially the well-known movement NBA, successfully stalled some of the major project works for long stretches of time. Besides, Madhya Pradesh has to go through judicial scrutiny in many cases filed against dam projects in the high court and Supreme Court. Thus, Madhya Pradesh has to complete a herculean task within the given timeline amid facing adversities.

Looking at the overall view of the Narmada basin in terms of utilization of Narmada water, the state of affairs is not quite satisfactory. Out of the total 34537.44 MCM (28 MAF), the calculated average availability of utilizable waters in the basin live storage capacity created through dam reservoirs till the year 2014–15 is 22,070.24 MCM (17.89 MAF). The water withdrawal by the party states for utilization in irrigation, domestic, and industrial purposes is also trailing behind. From 2006–07 till 2014–15 water years (July to June), a total of 14,955.70 MCM (12.12 MAF) water is being utilized each year. Madhya Pradesh is utilizing 7,922.04 MCM (6.42 MAF), Gujarat 6,520.66 MCM (5.29 MAF), and Rajasthan 513 MCM (0.41 MAF) water in each water year, that is, from 2006–07 to 2014–15.

Madhya Pradesh, which has been facing adversities till now, succeeded in creating water storage capacity of 16,902 MCM (13.70 MAF) through major, medium, and minor dam projects. The utilization of Narmada water is likely to increase in the state as the department of NVD has taken up projects for the interbasin use of

Narmada water. These projects are being designed, keeping in view the minimum land acquisition and loss of forest and environment.

It is important to know that the construction of dam irrigation projects alone does not guarantee the utilization of shared water. The party states will have to develop the command areas of the irrigation projects. The scientific development of command areas will ensure the ultimate goal of water utilization. Besides, the capacities of farmers will have to be built to manage the flow irrigation system, training them for participatory irrigation management and adopting advance farming practices. With this, the states will also have to ensure the availability of all necessary inputs of advance farming like seeds, fertilizers, and farming equipments. The infrastructure for storage of agriculture produces and proper marketing facilities would also add to the maximum utilization of the available Narmada water.

NARMADA–NEW PLACE OF PRIDE AND HOPE

Amidst disputes and hurdles, Narmada is all set to reach its innumerable destinations before its final submission into the Arabian Sea. The river and its tributaries are now on the right path to reach far-off places in the valley. The places where lakhs of people are reeling under poverty and poverty-related hardships of life. With their small pieces of land, and dependent on dry farming, they had never imagined that there would be canals across their fields full of water, and they would able to multiply agricultural production from the same small piece of land through irrigation.

Crossing every hurdle, the major dam irrigation projects on the eight tributaries of Narmada in Madhya Pradesh, that is, Maan, Hathni, Tawa, Barna, Kolar, Sukta, Matiyari, and Upper Beda have been completed. A major lift irrigation project (Punasa Lift), lifting water from Indira Sagar reservoir, has also been completed. All these projects provide irrigation to their large command areas. The three other dams of major irrigation projects on Narmada named Rani Avanti Bai Sagar, Indira Sagar, and Omkareshwar have been completed. The construction of canal networks of these projects is progressing. The irrigation benefits from these ongoing projects have also begun in a phased manner. The Sardar Sarovar Dam in Gujarat is on its way to achieve its full height after getting clearance to install gates on the crest of the dam. The main canal from this dam has been completed till the border of Rajasthan. From this point, Rajasthan has constructed it to further extend it to the state. The distribution network in Gujarat and Rajasthan is taking place speedily.

The irrigation potential in Madhya Pradesh, Gujarat, and Rajasthan is increasing every year as projects are being completed and ongoing projects are heading towards completion. The NWDT had put the proviso with the allocation of water to the states that the utilization of the water be assured till the year 2024. It is believed that by this timeline, Madhya Pradesh will achieve a target of creating 27.50 lakh hectare irrigation with the utilization of allocated 18.25 MAF water. Similarly, Gujarat and Rajasthan too will achieve the target of creating 18.62 lakh hectare and 2.46 lakh hectare respectively by utilizing 9 MAF and 0.50 MAF. Thus, in years to come, there will be 48.58 lakh hectare additional irrigation potential to bring the revolution in the agriculture of the Narmada Valley. The agricultural development will spark all other indicators of development, which will lead to the holistic development of Narmada Valley.

Narmada Valley dam irrigation projects are instrumental in giving a permanent solution to drinking water problems in the remote areas of the valley. Sardar Sarovar Project is making drinking water available to nearly 2,000 villagers and 54 urban areas in eight districts of Gujarat. Rajasthan has made provision to supply 0.106 MAF of water for drinking purposes in Jallore, Sanchor, Bhinmal, and 697 villagers of Jallore district. A total of 639 villagers of district Barmer would also get drinking water from Narmada canal in Rajasthan. It is estimated that nearly 15 lakh population in the towns and villages of Rajasthan would get a permanent solution to the age-old drinking water problem.

Madhya Pradesh has made provision to channel 937.74 MCM of water for drinking purposes to cities, towns, and villages. The state has come forward successfully for the transbasin use of water. Narmada-Kshipra Simhastha Link is one such project that is taking 5 cumecs of

Narmada water to the Kshipra basin of Malwa to provide drinking and industrial water, which is much needed there.

In addition to providing flow irrigation to a very large area under agriculture, Narmada Valley hydroelectric projects also generate power. This power is cheaper than other sources of power generation. Sardar Sarovar has the installed capacity of 1,450 megawatt, while the Indira Sagar and Omkareshwar projects have the installed capacity of 1,000 and 520 megawatt respectively. Maheshwar Hydroelectric Project has been completed with the installed capacity of 400 megawatt. Rani Avanti Bai Sagar Project is providing electric power with the installed capacity of 100 megawatt. There has been a continuous exercise to exploit the remaining possibilities of hydropower from Narmada and its tributaries. This additional hydropower would enhance the industrial activities in the valley. Thus, the man-made reservoirs in the valley provide yet another benefit of pisciculture activities.

Narmada has geared up for new socio-economic changes in the valley. The water is all set to reach millions of people of the valley to bless them with sustainable assets of a happier life. The river, with its place of pride and hope in the religious beliefs of millions of people for heavenly betterment, is now manifesting to ensure their worldly betterment too.

CHRONOLOGY OF EVENTS

1901: The first irrigation commission of India in its report observed the desirability of constructing a barrage on Narmada near Bharuch.

1946: Bombay and the Central Provinces and Berar requested the Central Waterways, Irrigation, and Navigation Commission (CWINC) to undertake investigations on Narmada with regard to flood control, navigation, irrigation, and hydroelectricity.

1948: The committee constituted under the chairmanship of A.N. Khosla recommended detailed survey and investigation of Bargi, Tawa, Punasa, and Bharuch projects.

1955: Central Water and Power Commission (CWPC) studied the hydroelectricity generation capacity of Narmada basin.

1957: In the meeting of the states of Madhya Pradesh and Bombay held under the chairmanship of the CWPC, a decision was taken to explore the possibilities of irrigation and hydroelectricity between Punasa and Bharuch. Dr K.L. Rao, the then member of CWPC, inspected in February the proposed site of Bharuch Project at village Gora.

1959: On the recommendation of Dr K.L. Rao, the CWPC shifted the site of Bharuch Project from Gora to a site of 2.5 kms upstream near Navagam (the present site of Sardar Sarovar Project). The dam at the new site was proposed in two phases. The Ministry of Irrigation and Power appointed a panel of experts in January to review the Bharuch irrigation project. The panel suggested the integration of the two phases of the proposed dam at Navagam with a height of 320 ft.

1960: Bombay was divided into two states—Maharashtra and Gujarat—under the States Reorganization Act, 1956 on 1 May 1960. The entire work relating to the proposed Navagam Dam was transferred to Gujarat. The construction of Navagam in two phases was approved by the Planning Commission in August. Approval was given for the generation of 625 megawatt hydroelectricity from the water released after meeting the irrigation requirements of Madhya Pradesh from Punasa (now Indira Sagar Project), and other dams constructed upstream of Navagam Dam.

1961: Administrative sanction by Gujarat in February for the construction of Phase I of Navagam Dam in the light of approval by the Planning Commission. Pandit Jawaharlal Nehru, the then prime minister of India, laid the foundation of Navagam Dam on 5 April.

1963: Agreement on draft prepared in Delhi regarding the height of Navagam Dam in a meeting at Bhopal between the Central Minister for Irrigation and Power and the chief minister of Gujarat and Madhya Pradesh. This was known as the Bhopal Agreement.

On 28 November, D.P. Mishra, the then chief minister of Madhya Pradesh, wrote a letter to Dr K.L. Rao, the then Central Minister of Irrigation and Power, conveying the decision of the government of Madhya Pradesh of the non-acceptance of Bhopal Agreement.

1964: Government of India constituted a high-level committee under the chairmanship of Dr A.N. Khosla, the then governor of Orissa, on 5 September.

1965: The Khosla Committee submitted its report to the government on 1 September.

1968: The complaint about Narmada water dispute was sent to the the Central Government by Gujarat in the month of July.

1969: Government of India constituted Narmada Water Dispute Tribunal (NWDT) on 6 October under section 4(1) of the Interstate Water Dispute Act, 1956.

1972: The NWDT pronounced a decision on the legal issues related to the dispute in July. A master plan regarding the construction of 29 major, 135 medium, and about 3,000 minor dams was submitted to the NWDT by Madhya Pradesh.

1972: On 22 July, the states of Madhya Pradesh, Gujarat, Maharashtra, and Rajasthan reached an agreement to allot 0.5 MAF water to Rajasthan. It was also agreed that the dispute about Narmada was to be submitted to the prime minister instead of the tribunal.

1974: On 12 July, the four states struck an agreement. Under the agreement, Madhya Pradesh and Rajasthan agreed to withdraw the petitions from the Supreme Court and decided to entrust the Narmada water dispute again to the tribunal.

1978: The NWDT released analytical details in three volumes about the issues pertaining to the water dispute, along with report about conclusions.

1979: On 7 December, the NWDT delivered a final decision on the Narmada water dispute. This decision was published on 12 December in the official gazette of Government of India.

1980: On 10 September, the Ministry of Irrigation constituted the Narmada Control Authority (NCA) and a review committee of the authority. The NCA started work from 20 December. Government of India constituted the Sardar Sarovar Construction Advisory Committee (SSCAC), and Gujarat constituted the Narmada Planning Group (NPG).

1984: On 23 October, Smt. Indira Gandhi, the then prime minister of India, laid the foundation of Narmada Sagar (now Indira Sagar) in Khandwa, Madhya Pradesh.

1985: On 16 July, Madhya Pradesh constituted Narmada Valley Development Authority (NVDA).

1987: Environmental sanction from the Ministry of Environment and Forests was received in respect of Sardar Sarovar and Indira Sagar projects on 24 June. Work on the Sardar Sarovar Dam was commenced.

1988: Sardar Sarovar Narmada Nigam Ltd was incorporated in Gujarat.

1989: In Madhya Pradesh, the construction of Rani Avanti Bai Sagar major dam on Narmada was completed. In Gujarat, a massive rally opposing the dam was organized at Sardar Sarovar Dam.

1992: The Morse Committee of the World Bank submitted its report on the Sardar Sarovar Project before the World Bank. In Madhya Pradesh, the construction of Maheshwar Hydroelectricity Project on Narmada was entrusted to Maheshwar Hydel Power Corporation Ltd.

1994: On 17 March, a four-member team of the Ministry of Environment and Forests visited the Sardar Sarovar Dam. The NBA filed a PIL, seeking review and stay on the entire work of the dam. The PIL was admitted by the Supreme Court.

2000: The height of the Sardar Sarovar Dam was raised to 90 metres. The government of Madhya Pradesh and National Hydroelectric Power Corporation (NHPC) incorporated the Narmada Hydroelectric Development Corporation (NHDC). On 18 October came the historic judgment of the Supreme Court, permitting the construction of Sardar Sarovar Dam to the proposed full height of 138.68 metres.

2001: In March, for the first time, water from Sardar Sarovar Dam was released in the main Narmada canal.

2002: In July, the height of the Sardar Sarovar Dam was raised to 95 metres.

2003: In July, the height of the Sardar Sarovar Dam was raised to 100 metres. On 30 August, Atal Bihari Vajpayee, the then prime minister of India, laid the foundation of Omkareshwar Dam.

2004: In June, the height of the Sardar Sarovar Dam was raised to 110.64 metres. The first unit of Indira Sagar Project started the generation of power. The town of Harsud, facing the threat of submergence from Indira Sagar reservoir, was evacuated in June. In August, electricity generation started from the canal powerhouse of Sardar Sarovar Project.

2005: In February, the first unit of riverbed powerhouse of Sardar Sarovar Project started generation of electricity. The construction of Indira Sagar Dam was completed. On 15 March, Supreme Court gave an important direction of allotting a minimum of two hectares of land to the adult son in a family affected by the Sardar Sarovar Project.

2006: In Madhya Pradesh, the construction of a dam on Maan, a tributary of Narmada, was completed. The project was dedicated to the public on 5 October by L.K. Advani, the then deputy prime minister of India. In April, the government appointed Shunglu Committee to check the rehabilitation work pertaining to Sardar Sarovar Dam in Madhya Pradesh. In December, the dam was raised to 121.92 metres.

2007: In Gujarat, 1,450 megawatt powerhouse of the Sardar Sarovar Dam was dedicated to the nation in a ceremony by the chief minister of Gujarat Narendra Modi. In Madhya Pradesh, electricity generation from Omkareshwar Project was initiated.

2008: The main canal of Sardar Sarovar Dam was completed, and Narmada water started flowing to Rajasthan. Jobat Dam on Hathni river, a tributary of Narmada, was completed.

2009: In order to focus attention on the pending medium and minor projects in Narmada Valley, the government of Madhya Pradesh constituted the Narmada Basin (medium and minor irrigation) Project Company Ltd (NBPCL). The NVDA and the NHDC entered into an agreement on 29 June to construct 1,230 megawatt thermal power project (Rewa Thermal Power Project) in Khandwa.

2010: The stay granted by the Madhya Pradesh High Court on the petition of the NBA against the construction of canals of Indira Sagar and Omkareshwar projects was vacated in February by the Supreme Court with certain conditions. Farmers losing more than 60% of their holdings in the construction of canals of Indira Sagar and Omkareshwar dams were to get land as per the order of the Supreme Court passed in May.

2011: The appeal by the government of Madhya Pradesh against the order of the high court regarding Omkareshwar Project. The Supreme Court held on 11 May that the provisions of the NWDT award do not apply to other projects, including Omkareshwar Project in Madhya Pradesh. In the same case, the Supreme Court made scathing comments against the NBA for furnishing false information in the affidavit.

2012: The Narmada Control Board (NCB) meeting was chaired by the chief minister of Madhya Pradesh Shivraj Singh Chouhan, who accorded clearance for the Narmada-Kshipra

Link Project aimed at providing drinking and industrial water to Malwa.

2013: Madhya Pradesh declared special rehabilitation package on 28 May for the Omkareshwar reservoir-affected families. The package amount is estimated to be Rs. 212.87 crores.

2014: The Narmada-Kshipra Simhastha Link Project in Madhya Pradesh was completed. Former deputy prime minister L.K. Adwani and chief minister Shivraj Singh Chouhan dedicated the project to the people on 25 February by discharging Narmada water into the origin of the river Kshipra at a function.

2014: The NCA accorded sanction to install gates at the crest level of Sardar Sarovar Dam on 12 June. The matter was pending for many years. It was taken on priority after the new central government took charge under the leadership of the former Gujarat chief minister Narendra Modi as the prime minister of India. The government of Gujarat welcomed the decision, and the work for the installation of the gates started the following day.

2014: Narmada Bachao Andolan (NBA) registered objection on the decision of raising the height of the Sardar Sarovar Dam in Madhya Pradesh High Court on 8 July 2014, pleading that since the Jha Commission was yet to submit its report on various issues of rehabilitation, the decision to raise the dam height was not justified. On the issue, the NBA submitted transfer petition for transferring the case from Madhya Pradesh High Court to Supreme Court on 21 July 2014. In connection to the NCA petition, the Supreme Court clarified on 4 August that the issue related to the dam height would be heard only by it.

2015: The government of Madhya Pradesh accorded sanction for the construction of Narmada-Gambhir River Link project that is designed to create 50,000 hectares of irrigation potential in Malwa region.

2016: Madhya Pradesh dropped the proposed dam irrigation projects, namely Chinki, Ataria, Upper Budhner, Sher, Machhrewa, and Sitarewa. This move came in view of the issues that cropped up due to the new land acquisition act enforced from 2014.

2017: Madhya Pradesh shifted its approach from dam irrigation projects to lift irrigation. As much as sixteen lift irrigation projects on canals and reservoir accorded clearance to speed up the utilization of allocated water.

2017: The Supreme Court bench headed by Chief Justice J.S. Khehar, on 8 February, disposed all petitions on land for the land issue of Sardar Sarovar Dam in Madhya Pradesh. Quoting a full and final settlement, the court ordered to pay Rs. 60,00,000 to each family who had not accepted the package of the government. It also ordered to pay Rs. 15,00,000 to each family who are identified by the Jha Commission as being exploited by middlemen. The Supreme Court also fixed the timeline to vacate the reservoir area before 31 July 2017.

2017: On 5 June, Madhya Pradesh government declared a package amounting to Rs. 900 crores for the dam-afflicted families.

This package was for additional relief to families not covered under Supreme Court order dated 8 February 2017.

2017: On 31 July, the Supreme Court rejected the petition of the NBA to pass a stay on the shifting of families and extend the timeline.

2017: Medha Patkar, while sitting on hunger strike at village Chikhalda (Dhar), was forcefully shifted to Indore on 7 August by the administration. She got admitted to a local Bombay hospital for treatment and was released on 9 August. The same day, she got arrested by the police while heading towards Dhar to meet the dam oustees.

2017: Medha Patkar continued hunger strike in the district jail of Dhar. She broke her fast on 12 August when the social worker Chinmaya Mishra reached Dhar jail and offered her lime water. She was released on bail on 24 August.

2017: On 25 August, the Supreme Court dismissed a contempt petition filed by the NBA against the Madhya Pradesh government and PIL by some affected families in connection with the shifting of people falling within the region of Sardar Sarovar. Medha Patkar organized a journey for justice (Nyay Yatra) in the affected villages, which started from Rajghat (Barwani) on 30 August.

2017: Almost all families residing in the reservoir area shifted till the first week of September. Gujarat started raising the water level after closing the gates. Prime Minister Narendra Modi inaugurated the Sardar Sarovar Dam at a function on 17

September in the village Kevadiya. Hours later, Medha Patkar suspended the jal satyagraha she had started at the village Chhota Badada in Madhya Pradesh.

2018: Madhya Pradesh succeeded in completing Narmada-Malwa-Gambhir Link Project, which is second in line in linking Narmada to four major rivers of Malwa region. About 15 cumecs of Narmada water has been lifted and taken to Gambhir river basin to irrigate 50,000 hectares of land. The work on the first phase of Narmada-Parvati Link has begun to creat 1 lakh hectares of irrigation potential in Malwa, Madhya Pradesh.

2019: The work on the first phase of Narmada-Kalisindh Link has begun to creat 1 lakh hectares of irrigation potential in Malwa, Madhya Pradesh.

ACKNOWLEDGEMENTS

While compiling this book, I have drawn references from various reports of Narmada Control Authority, Central Water Commission, and from publications of Gujarat and Madhya Pradesh governments. The chronology of events pertaining to Narmada river dispute and related political, social, and administrative activities has been drawn from reports that appear in newspapers like *The Hindu*, *Times of India*, *Hindustan Times*, *Dainik Bhaskar*, *Naidunia*, *Patrika*, etc. Peoples' struggle against big dams, submergence, displacement, and events leading to the emergence of Narmada Bachao Andolan are taken from various sources, including interactions with eyewitnesses. References have also been drawn from books titled *The River and Life: People's Struggle in the Narmada Valley* (1998) by Sanjay Sangvai and *Human Resettlement in Lower Narmada Basin* (1985) by S.C. Verma.

I am grateful to the experts of various fields who have helped in refining the text, especially to the irrigation engineer L.K. Wagh. He was kind enough to guide me on technical matters and provide the chronology of events during the exercise of Narmada Water Dispute Tribunal. As a member of the team who helped the senior advocates plead on behalf of Madhya Pradesh, he was an eyewitness to the whole exercise. I express my sincere thanks for the unconditional support extended by the former IAS officer J.P. Vyas in bringing out this book.

For sharing his precious feedback, reading the manuscript, and lending his support to publish it, I would like to acknowledge the vital role played by journalist-cum-author, Atul K. Thakur. I am also thankful to journalist Rasheed Kidwai for helping me in my work.

Finally, I humbly dedicate this whole work to my mother Rabiya Mustafa Khan.

INDEX